Cancer and Vitamin C

OXIDATIVE STRESS AND DISEASE

Series Editors
Lester Packer, PhD
Enrique Cadenas, MD, PhD
University of Southern California School of Pharmacy
Los Angeles, California

Adipose Tissue and Inflammation, *edited by Atif B. Awad and Peter G. Bradford*

Herbal Medicine: Biomolecular and Clinical Aspects, Second Edition, *edited by Iris F.F. Benzie and Sissi Wachtel-Galor*

Inflammation, Lifestyle, and Chronic Diseases: The Silent Link, *edited by Bharat B. Aggarwal, Sunil Krishnan, and Sushovan Guha*

Flavonoids and Related Compounds: Bioavailability and Function, *edited by Jeremy P.E. Spencer and Alan Crozier*

Mitochondrial Signaling in Health and Disease, *edited by Sten Orrenius, Lester Packer, and Enrique Cadenas*

Vitamin D: Oxidative Stress, Immunity, and Aging, *edited by Adrian F. Gombart*

Carotenoids and Vitamin A in Translational Medicine, *edited by Olaf Sommerburg, Werner Siems, and Klaus Kraemer*

Hormesis in Health and Disease, *edited by Suresh I.S. Rattan and Éric Le Bourg*

Liver Metabolism and Fatty Liver Disease, *edited by Oren Tirosh*

Nutrition and Epigenetics, *edited by Emily Ho and Frederick Domann*

Lipid Oxidation in Health and Disease, *edited by Corinne M. Spickett and Henry Jay Forman*

Diversity of Selenium Functions in Health and Disease, *edited by Regina Brigelius-Flohé and Helmut Sies*

Mitochondria in Liver Disease, *edited by Derick Han and Neil Kaplowitz*

Fetal and Early Postnatal Programming and Its Influence on Adult Health, *edited by Mulchand S. Patel and Jens H. Nielsen*

Biomedical Application of Nanoparticles, *edited by Bertrand Rihn*

The Biology of the First 1,000 Days, *edited by Crystal D. Karakochuk, Kyly C. Whitfield, Tim J. Green, and Klaus Kraemer*

Hydrogen Peroxide Metabolism in Health and Disease, *edited by Margreet C.M. Vissers, Mark Hampton, and Anthony J. Kettle*

Glutathione, *edited by Leopold Flohé*

Vitamin C: New Biochemical and Functional Insights, *edited by Qi Chen and Margreet C.M. Vissers*

Cancer and Vitamin C, *edited by Qi Chen and Margreet C.M. Vissers*

For more information about this series, please visit:
https://www.crcpress.com/Oxidative-Stress-and-Disease/book-series/CRCOXISTRDIS

Cancer and Vitamin C

Edited by
Qi Chen, PhD
Associate Professor
Department of Pharmacology, Toxicology and Therapeutics
University of Kansas Medical Center
Kansas City, Kansas, USA
Margreet C.M. Vissers, PhD
Research Professor
Department of Pathology and Biomedical Science
University of Otago, Christchurch, New Zealand

CRC Press is an imprint of the
Taylor & Francis Group, an **informa** business

CRC Press
Taylor & Francis Group
6000 Broken Sound Parkway NW, Suite 300
Boca Raton, FL 33487-2742

Printed on acid-free paper

International Standard Book Number-13: 978-0-367-85804-9 (Hardback)

Visit the Taylor & Francis Web site at
http://www.taylorandfrancis.com

and the CRC Press Web site at
http://www.crcpress.com

DEDICATION

This book is dedicated to the memory of

Lester Packer
August 28, 1929–July 27, 2018

With thanks for his tireless work and his endless enthusiasm for science, and in particular, for his work as editor of the Oxidative Stress and Disease series of books. This project (now two volumes) was initiated on Prof Packer's invitation at the end of 2017.

CONTENTS

PREFACE

Since its discovery almost a century ago as the ingredient responsible for the antiscorbutic properties of fruits and vegetables, there has been interest in the potential for vitamin C as a remedy for many illnesses. However, no research area has created as much controversy as the investigations on whether vitamin C has a role to play in the treatment or management of cancer. This topic has received widespread attention from clinicians, scientists, and the general public since the clinical experiments of Cameron and Pauling were reported in the 1970s. They reported that high doses of intravenous and oral ascorbate administered to patients with advanced cancer could improve patient survival times and even induce complete remissions. However, these claims were met with widespread incredulity, mostly because there was no plausible mechanism of action to account for any anticancer mechanism. Famously, Cameron and Pauling's studies were brought into question when a subsequent randomized and placebo-controlled study from the Mayo Clinic found no significant benefit for cancer patients when ascorbate was given orally. However, despite the clinical uncertainty, high-dose vitamin C generally by infusion continued to be a commonly sought treatment option from complementary and alternative medicine providers. Case studies continued to surface with suggestions that there may be circumstances under which vitamin C could provide a clinical benefit, and it was noted that the treatment usually involved high intravenous doses, given as a course over months.

During the past 15 years, there has been a resurgence of interest in the potential for vitamin C as an anticancer treatment. This renewed attention has followed more in-depth studies of the pharmacokinetics of vitamin C, coupled with consideration of its biochemistry and redox properties. The aim of this book is to provide an up-to-date overview of the significant advances in the study of vitamin C and its role in the cancer clinic. The contents of this book are written against a background of the information provided in the first of these companion books, *Vitamin C: New Biochemical and Functional Insights*. Together, these two books provide a comprehensive overview of the current expanded understanding of the many roles vitamin C plays throughout the body. Much new information is included. The last books on vitamin C were published in 1999 and 2003, and with renewed interest in its potential clinical use in infectious disease, in neurological and psychiatric illnesses, and in cancer, we believe the time is right for these publications.

The first chapter of this book is a historical gem from Lawson, telling the story of the early studies into vitamin C on which much of the current interest is still based. There is still much to learn from these early observations. Today, misquoting of these early studies is common, and this chapter provides an appropriate in-depth account of this pioneering work and the context in which it arose. It behooves anyone interested in vitamin C and micronutrients and their role in medicine to read this chapter.

A more sophisticated understanding of the biochemistry of vitamin C has led to the emergence of several possible theories that might explain how high-dose vitamin C could influence tumor growth and survival. These are detailed in Part II. Interest in vitamin C was renewed following demonstrations that concentrations achievable by intravenous administration are selectively toxic to cancer cells *in vitro*. This activity is due to the pro-oxidant activity of vitamin C, resulting in significant production of H_2O_2 at these high concentrations. This work is elegantly summarized, and functional targets in cancer cells are identified in the chapters by Chen et al.

Many studies have demonstrated that vitamin C administration (by oral, intravenous, or intraperitoneal route) can significantly reduce tumor growth rates in mice. In some of these studies, this is associated with the ability of ascorbate to modulate tumor hypoxia and inhibit angiogenesis via inhibition of the pro-survival transcription factor hypoxia-inducible factors (HIFs). The HIFs promote the growth and survival of solid tumors, and vitamin C is a cofactor for the regulatory hydroxylases that control this transcription factor activation. The current state of knowledge on this topic and the evidence that vitamin C availability could impact on HIF activation in developing tumors are described in Chapter 4 by Wohlrab, Kuiper, and Dachs.

A substantial chapter by Cullen and Alexander describes the potential for high-dose vitamin C to act in concert with radiation treatment. The potential for the antioxidant activity of vitamin C to counter the clinical benefit of established cancer therapies that depend on pro-oxidant activity is a legitimate clinical concern. Radiation treatment is the best exemplar of a pro-oxidant anticancer treatment, and information that will enhance our understanding of the potential interactions with vitamin C will be welcomed by clinicians.

It is now recognized that the establishment of a cancer cell phenotype is accompanied by epigenetic dysregulation. Identification of the ascorbate-dependent enzymes responsible for histone and DNA demethylation has significantly expanded the scope for vitamin C to have impact on cancer biology. This exciting area of research is described in the chapter by Wang and Cimmino on the regulation of somatic cell reprogramming and the influence of vitamin C on the enzymes responsible for these reactions. This research is of particular interest for the hematologic cancers as described.

Together, these proposed functions have stimulated clinical studies, with a significant number already completed or underway. These are described in the very useful and extensive summaries in the chapters by Paller et al. and Drisko.

As with the first book, we thank the contributors of this book for their generosity and enthusiasm in the preparation of the chapters. All are highly committed and busy individuals, and their contributions are invaluable. The aim of both books is to provide an overview of the current state of knowledge that reflects the renewed interest in vitamin C. We trust that the content of this book, and its companion book on the new biochemical and functional insights of vitamin C, will be of interest to research scientists, clinicians, students of medical science and nutrition, interested patient groups, and general readers.

QI CHEN
MARGREET C.M. VISSERS

CONTRIBUTORS

MATTHEW S. ALEXANDER
Free Radical and Radiation Biology Program
Departments of Surgery and Radiation Oncology
Holden Comprehensive Cancer Center
University of Iowa College of Medicine
Iowa City Veterans Affairs
Iowa City, Iowa

RAMESH BALUSU
Department of Internal Medicine
Division of Hematologic Malignancies and Cellular Therapeutics
University of Kansas Medical Center
Kansas City, Kansas

PING CHEN
Department of Pharmacology, Toxicology, and Therapeutics
University of Kansas Medical Center
Kansas City, Kansas

QI CHEN
Department of Pharmacology, Toxicology, and Therapeutics
University of Kansas Medical Center
Kansas City, Kansas

LUISA CIMMINO
Department of Biochemistry and Molecular Biology
Miller School of Medicine
University of Miami
Miami, Florida

JOSEPH J. CULLEN
Free Radical and Radiation Biology Program
Departments of Surgery and Radiation Oncology
Holden Comprehensive Cancer Center
University of Iowa College of Medicine
Iowa City Veterans Affairs
Iowa City, Iowa

GABI U. DACHS
Mackenzie Cancer Research Group
Department of Pathology and Biomedical Science
University of Otago Christchurch
Christchurch, New Zealand

RUOCHEN DONG
Department of Pharmacology, Toxicology, and Therapeutics
University of Kansas Medical Center
Kansas City, Kansas

JEANNE A. DRISKO
Department of Internal Medicine
Integrative Medicine Program
University of Kansas Medical Center
Kansas City, Kansas

AMY GRAVELL
The Emmes Corporation
Rockville, Maryland

CAROLINE KUIPER
Centre for Free Radical Research
Department of Pathology and Biomedical Science
University of Otago Christchurch
Christchurch, New Zealand

STEPHEN LAWSON
Linus Pauling Institute
Oregon State University
Corvallis, Oregon

MARK LEVINE
National Institute of Diabetes and Digestive and Kidney Diseases
National Institutes of Health
Bethesda, Maryland

THOMAS LUECHTEFELD
Bloomberg School of Public Health
Johns Hopkins University
Baltimore, Maryland

CHANNING PALLER
Sidney Kimmel Comprehensive Cancer Center
Johns Hopkins University
Baltimore, Maryland

KISHORE POLIREDDY
Department of Pharmacology, Toxicology, and Therapeutics
University of Kansas Medical Center
Kansas City, Kansas

TAMI TAMASHIRO
The Emmes Corporation
Rockville, Maryland

JINGJING WANG
Department of Pathology
NYU School of Medicine
New York, New York

TAO WANG
Department of Pharmacology, Toxicology, and Therapeutics
University of Kansas Medical Center
Kansas City, Kansas

CHRISTINA WOHLRAB
Mackenzie Cancer Research Group
Department of Pathology and Biomedical Science
University of Otago Christchurch
Christchurch, New Zealand

PART I

Historical Overview of Cancer and Vitamin C

CHAPTER ONE

Introduction to Vitamin C and Cancer Clinical Studies

A HISTORICAL PERSPECTIVE

Stephen Lawson

CONTENTS

INTRODUCTION

The clinical use of vitamin C as adjunctive therapy for cancer has a long and controversial history. Its most celebrated advocate, Linus Pauling, was one of the twentieth century's most important and influential scientists and the only person to have won two unshared Nobel Prizes (chemistry, 1954, and peace, 1962). He won the Nobel Prize in chemistry for elucidating the nature of the chemical bond and its application to understanding the structure of complex substances. Pauling's concept of biological specificity and his discovery of the structural themes of protein molecules, including the α helix, laid the foundation for the science of molecular biology. In 1949, Pauling published a seminal paper in *Science* that reported the molecular cause of sickle cell anemia—the first disease to be characterized as a molecular disease—thereby heralding the era of molecular medicine. In the early 1960s, Pauling and Zuckerkandl proposed that hemoglobin could be used as an evolutionary clock to date the divergence of species—a concept that led to the science of molecular evolution. In the mid-1960s, Pauling became interested in the role of micronutrients in treating mental illness and began to focus on the prophylactic and therapeutic functions of vitamin C. He termed this approach "orthomolecular medicine"—using micronutrients in the right amounts to prevent and treat disease. This interest led to lengthy collaborations with several clinicians, including Ewan Cameron in the early 1970s and Abram Hoffer, on the use of high-dose vitamin C in cancer. Recent work on the pharmacokinetics of vitamin C and on the anticancer molecular mechanisms of vitamin C have inspired the resurgence of interest in clinical applications. The histories of the early studies that stimulated this contemporary research revival are outlined in this chapter.

ORTHOMOLECULAR MEDICINE

In 1968, *Science* published an article by Linus Pauling entitled, "Orthomolecular Psychiatry," [1] in which Pauling outlined the evidence to support his contention that "Biochemical and genetic arguments support the idea that orthomolecular therapy, the provision for the individual person of the optimum concentrations of important normal constituents of the brain, may be the preferred treatment for many mentally ill patients." In his discussion of the mental symptoms that accompany B-vitamin deficiencies, he noted that Norwegian investigators had found abnormally low levels of vitamin B_{12} in the blood of about 15% of patients admitted to a mental hospital and that niacin had been used successfully to treat pellagra-associated psychosis in the southeastern United States in the early twentieth century. He also cited the therapeutic use of high-dose niacin (0.3–18 g/d) by a number of researchers, including Cleckley, Sydenstricker, and Hoffer, to treat psychiatric illnesses like schizophrenia, and briefly discussed Stone's contention, based on cross-species comparisons and other arguments, that the optimum intake of vitamin C might be 3–15 g/d. Pauling discussed the loss of the synthetic ability of vitamin C in a few mammalian species about 20 million years ago and suggested that such a mutation would have been beneficial in an environment supplying sufficient vitamin C exogenously. Over time, the mutant would have an evolutionary advantage in applying energy saved from synthesizing the optimum amount of the vitamin to other functions.

Building on his previous work on the molecular basis of sickle cell anemia, Pauling described phenylketonuria as a molecular disease, caused by a defect in the gene coding for phenylalanine hydrolase that catalyzes the conversion of phenylalanine to tyrosine, which can be attenuated by the dietary restriction of phenylalanine-containing food. Pauling also noted that large doses of L-glutamic acid (from two to four times the amount found in food) had been successfully used in the 1940s to treat patients with mental retardation or convulsive disorders. Here, Pauling was arguing for the optimum concentration in the brain of biochemicals that affect function. In some cases, certain biochemicals like B vitamins may need to be increased, while others may need to be limited. Since the rates of enzyme-catalyzed reactions are proportional to the concentration of the reactant, he postulated that, in the case of defective enzymes or deficiencies in their amounts, the rates could be normalized by increasing the concentration of the substrate. Hence, high-dose B vitamin therapy would be expected to be therapeutic if such enzyme disorders occurred in the brain. More recently, Ames et al. discussed the remediation of about 50 genetic diseases caused by defective enzymes with high-dose B vitamins and other micronutrients [2].

Pauling applied a model of fluid dynamics to reservoirs in the body, suggesting that localized deficiencies of vital substances could occur in

Linus Pauling with molecular models (ca. early 1960s, from the Ava Helen and Linus Pauling Papers, OSU Libraries Special Collections & Archives Research Center).

some compartments, such as the cerebrospinal fluid, while normal concentrations would be observed in the blood. Therefore, the steady-state concentration of a vitamin in the brain could be less than that in the blood. This situation could be aggravated by genetic abnormalities in schizophrenics that might require very large doses of certain vitamins to rectify.

Another line of reasoning Pauling emphasized concerned bacterial genetics. He cited the *Neurospora* research by Beadle and Tatum showing that the growth rate of a pyridoxine-requiring mutant could be increased by about 7% more than that of the parental strain by supplying large amounts of pyridoxine in the growth media.

In subsequent publications, Pauling broadened the application of orthomolecular therapy to other aspects of medicine. In 1974, he defined orthomolecular medicine as "the preservation of health and the treatment of disease by the provision of the optimum molecular constitution of the body, especially the optimum concentration of substances that are normally present in the human body and are required for life. *The adjective orthomolecular is used to express the idea of the right molecules in the right concentration*" [3]. He further developed the importance of vitamin C in the prevention and treatment of infectious illnesses and introduced more examples of orthomolecular medicine, including the prevention of goiter by iodine and the treatment of diabetes mellitus by insulin. In Pauling's 1986 book for health professionals and the lay public, *How to Live Longer and Feel Better*, he mentioned two other diseases controlled by orthomolecular therapy: galactosemia, which is controlled by a galactose-free diet, and methylmalonicaciduria, which can be ameliorated in many patients by providing extremely high amounts of vitamin B_{12}, which normalize the rate of the enzymatic reaction that converts methylmalonic acid to succinic acid [4].

While maintaining interest in the role of assorted micronutrients in health, Pauling began to focus much attention on vitamin C. His 1970 bestselling book, *Vitamin C and the Common Cold*, outlined his arguments on the prophylactic and therapeutic value of high-dose vitamin C on colds [5]. Pauling reviewed the clinical literature, which was limited to four studies that had used 1 or more grams of vitamin C per day to assess the impact on colds. The publicity surrounding Pauling's work on vitamin C stimulated others to conduct controlled clinical trials of vitamin C and the common cold. By 1994, 21 such studies that used at least 1 g of vitamin C per day had been published, showing that vitamin C ameliorated symptoms and shortened the average duration of colds by about 23%, although there was no consistent effect on incidence [6]. Subsequent studies have generally supported that conclusion, especially for individuals under extreme physical stress.

In a 1970 paper [7], Pauling discussed the rationale for his conclusion that the optimum intake of vitamin C is at least 2.3 g/d. He analyzed the quantity of vitamin C, thiamin, riboflavin, and nicotinic acid in the amount of 110 raw plant foods, including nuts, fruit, grain, beans, and vegetables, that provides 2500 kcal of energy per day. The average values for thiamin, riboflavin, nicotinic acid, and vitamin C were 5.0, 5.4, 41, and 2300 mg, respectively. For the B vitamins, the ratio of these amounts to the recommended allowance at the time was about 3, whereas the ratio for vitamin C was 35. Pauling speculated that adaptive mechanisms, such as an increased ability of the kidneys to pump vitamin C back into the blood from the glomerular filtrate, storage of vitamin C in the adrenals, and/or the increased ability of cells to extract vitamin C from blood, allowed animals without the ability to synthesize vitamin C to thrive.

Several experiences in Pauling's life set the stage for his development of the concept of orthomolecular medicine. His mother suffered from psychosis ("megaloblastic madness") associated with pernicious anemia and died from the vitamin B_{12} deficiency disease in 1926, the same year that Minot and Murphy discovered that consuming raw liver—a good source of vitamin B_{12}—cured pernicious anemia. Pauling himself was diagnosed in 1941 with Bright's disease or glomerulonephritis—an incurable and frequently fatal disease at the time. Thomas Addis, his innovative physician, prescribed a diet low in salt and protein, supplementary vitamins and minerals, and plenty of water to rest the kidneys. Pauling followed the diet for about 14 years and completely recovered.

VITAMIN C AND CANCER

Ewan Cameron, chief of surgery at the 440-bed Vale of Leven Hospital in Loch Lomondside,

Ewan Cameron (ca. 1975).

Scotland, collaborated with Pauling on vitamin C and cancer studies for nearly 20 years beginning in 1971, 2 years before the founding of the Linus Pauling Institute of Science and Medicine in Menlo Park, California. Pauling was a chemistry professor at Stanford University and, as previously mentioned, had become interested in the prospect that large daily doses of vitamin C may help prevent chronic disease and infections and, in many cases, be useful in the clinical management of such diseases.

Although Pauling's initial interest in vitamin C was focused on its putative role in preventing and treating the common cold, influenza, and other infectious illness, he began to raise the possibility that the vitamin may also be important in reducing the risk for heart disease and cancer. In Scotland, Cameron read a news account in the *New York Times* of Pauling's comments on vitamin C and cancer and contacted him to discuss clinical applications [4]. Cameron had been interested in controlling cancer by inhibiting hyaluronidase, an enzyme secreted by proliferating cells to break down hyaluronic acid and other glycosaminoglycans in the intercellular matrix, thereby creating space for new cells, as outlined in his 1966 book *Hyaluronidase and Cancer* [8]. According to Cameron, the abnormally constant release of hyaluronidase by rapidly proliferating cancer cells is critical for their invasiveness. Cameron suspected that vitamin C might either serve as a physiological hyaluronidase inhibitor or boost the production of a hyaluronidase inhibitor, thereby retarding tumor growth. In 1954, McCormick had suggested that the depolymerization of glycoproteins, especially collagen, observed in cancer could be secondary to the vitamin C deficiency frequently observed in cancer patients [9], and in a 1959 paper [10], he presented more evidence from other investigators to support his hypothesis that such depolymerization might be inhibited by adequate vitamin C. Pauling noted that there was evidence that tumor cells also secrete collagenase to weaken collagen in the intercellular matrix and surmised that vitamin C might inhibit this process by stimulating the production of collagen fibers, leading to tumor encapsulation. Pauling suggested to Cameron that he give 10 g/d of vitamin C to his terminal cancer patients and observe the results. Over the next two decades, Cameron and Pauling published scores of papers and a book, *Cancer and Vitamin C* [11], on their joint work.

While this research attracted much public and professional attention, vitamin C had been used to treat cancer, somewhat sporadically, for many decades, and research using cell culture and animal models had also been conducted for many years. Several clinical studies were published in the German literature in the 1930s and 1940s, including one by Deucher in 1940 that reported a favorable response in patients receiving 1–4 g/d of vitamin C [12]. Other research in the 1940s and 1950s reported that daily doses of 1–2 g of vitamin C combined with large doses of vitamin A or with B vitamins or the injection of zinc-ascorbate led to tumor stasis or clinical improvement in many patients [13]. In 1954, Greer reported the control of chronic myeloid leukemia and symptomatic relief in a patient taking 24.5–48 g/d of oral vitamin C [14]. The case was unusual in that relapse occurred when the patient discontinued the regimen at his physician's direction. In 1968, Cheraskin et al. [15] reported a controlled clinical trial with 54 patients with cervical cancer, half of whom underwent diet therapy—a high-protein, low-refined carbohydrate diet supplemented with multivitamin-minerals and 750 mg/d of vitamin C (as reported by Cameron and Pauling [13])—for 1 week prior to undergoing radiotherapy and continuing for 3 weeks after radiation. The response to radiation of patients receiving the nutritional therapy was statistically significantly better than controls for those with stages I–III cancer, but not for those with stage IV cancer, which was a subgroup with only two patients. The average score, based on analysis of vaginal smears using the Graham method, for nutritionally treated patients was 97.5; the average score for controls was 63.3.

VALE OF LEVEN CLINICAL STUDIES

Cameron and Pauling were encouraged by these reports and speculated about the anticancer mechanisms of vitamin C, mainly focused on host resistance and the inhibition of invasiveness—cytostasis—rather than cytotoxicity. As mentioned, these postulated mechanisms included inhibition of hyaluronidase and collagenase, as well as the enhancement of immunocompetence through increased antibody and complement production and improved phagocytosis. As Cameron wrote to Pauling in 1972,

> I am convinced that [ascorbate] produces a favorable shift in the host/tumor relationship. I think that we are achieving tumor retardation and prolongation of life in many patients even although I fully appreciate that these factors are notoriously difficult to measure. What is perhaps

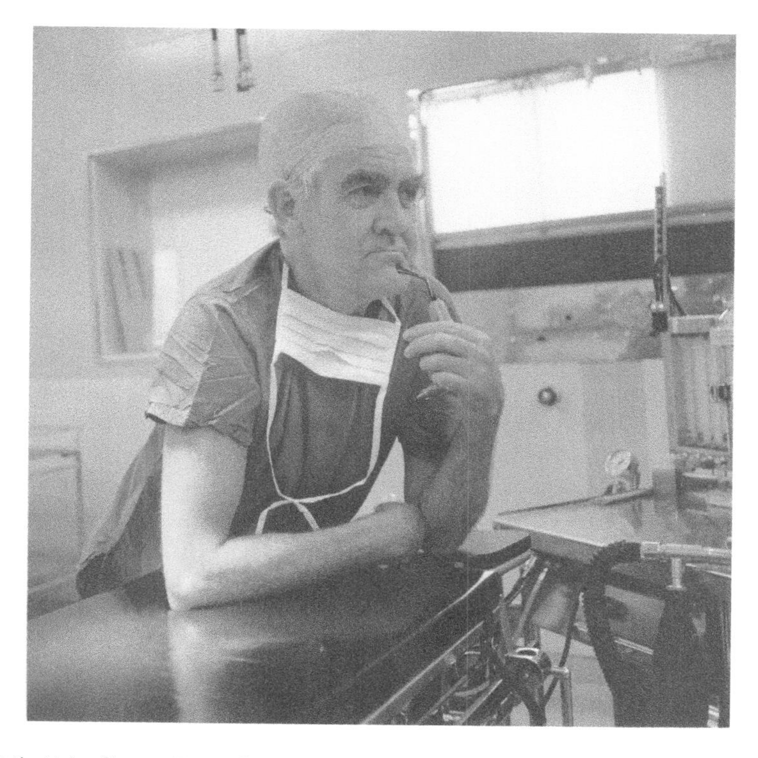

Ewan Cameron at the Vale of Leven Hospital in Scotland (ca. 1973).

more important than mere prolongation of life, is that by symptomatic relief, we are improving the quality of life [16].

Cameron and Pauling submitted a manuscript detailing their hypothesis about the putative anticancer effect of vitamin C and advocating therapeutic studies to the *Proceedings of the National Academy of Sciences*. As a member of the National Academy of Sciences, Pauling had the privilege of publication and was confounded when their paper was rejected. The editor explained that the paper would be better suited for publication in a medical journal. After its subsequent rejection by the *British Medical Journal*, the paper was published in *Oncology* in 1973 [16].

Pauling wanted to interest the National Cancer Institute (NCI) in conducting a randomized controlled trial (RCT) of vitamin C in advanced cancer and was invited to meet with NCI officials in 1973. At the meeting, Pauling presented the clinical results of Cameron's first 40 patients treated with high-dose vitamin C given by continuous slow-drip infusion and/or orally but was met with deep skepticism. The officials insisted that animal experiments needed to be conducted before such human trials could be conducted [11,16].

In 1974, Cameron and Pauling were invited to Memorial Sloan Kettering Cancer Center to present their hypothesis and discuss Cameron's clinical observations. Charles Young was impressed and decided to give 10 g/d of vitamin C by oral or intravenous administration to 23 patients with a variety of advanced cancers for 5–64 days. Young had previously observed transitory tumor regression in two patients with Hodgkin disease given high-dose vitamin C in addition to other anticancer therapy. He reported to Pauling in 1976 that, with the exception of one woman with diffuse histiocytic lymphoma (reticulum cell sarcoma) who had a minimal response, the patients had no obvious response to vitamin C, and the Memorial Sloan Kettering Cancer Center did not pursue further clinical investigations of vitamin C in cancer [17].

Cameron and Campbell published their cautious report on the first 50 consecutive patients with advanced cancer treated with high-dose vitamin C in 1974 [18]. These patients were classified as untreatable by conventional therapies, so vitamin C was the sole intervention, although some patients had been treated previously with surgery or radiation and a few with chemotherapy or hormones. The patients exhibited a variety of cancers, with most being cancers of the breast (7), bronchus (7), stomach (5), bladder (5), or colon (5). Metastases were common. Twenty-seven patients were female and 23 were male, ranging from 40 to 93 years old. Treatment with vitamin C was somewhat variable but typically consisted of 10 g/d infused intravenously for 2–11 days (mean = 7 days) and 10 g/d ingested orally for 2 to 1087 days or longer (mean = 186 days) either as the sole intervention or after intravenous infusion. A few patients received 45 g/d intravenously for up to 10 days. Tolerable side effects, including fluid retention and dyspepsia, were observed. Patient responses were categorized as no response (17), minimal response (10), growth retardation (11), cytostasis (3), tumor regression (5), or tumor hemorrhage and necrosis (4). The authors concluded that many of these patients survived longer than expected (although obviously difficult to determine objectively), experienced pain relief from skeletal metastases and relief from opiate dependence, and had reductions in hematuria and in the rate of malignant ascites and pleural effusions. Many patients also had reductions in erythrocyte sedimentation rates and in serum seromucoid concentrations, both markers for progressive cancer. Four patients had severe responses (tumor hemorrhage and necrosis), precipitating death 3–17 days after commencing vitamin C treatment.

Interestingly, several patients had responses to only orally administered vitamin C, usually 5 or 10 g/d for 1.5–661 days or longer. Eleven such patients had beneficial responses, including one with pancreatic cancer and one with kidney cancer, both of whom had tumor regressions. One of the aforementioned patients who succumbed to tumor hemorrhage and necrosis was treated only with oral vitamin C (10 g for 1.5 days). The observations that oral vitamin C without any intravenous administration seemed to provide benefit probably contributed to Cameron and Pauling's belief that intravenous vitamin C, while effective at raising blood concentrations, may not be essential in controlling cancer. Of course, the vitamin C transport system (SVCT1 and SVCT2) that governs and limits the absorption of vitamin C from the gut into the bloodstream was not discovered until 1999 [19], 5 years after Pauling's death. These few patients present a conundrum in the context of Levine's more recent work on the generation of selectively cytotoxic hydrogen peroxide by very high concentrations of extracellular vitamin C only attainable by intravenous administration of vitamin C [20]. In the case of Cameron's patients, the results may have been coincidental, which seems unlikely; other anticancer mechanisms may have been active; or there may have been something unusual about the patients' characteristics. For example, the Scottish diet at the time was not abundant in vitamin C, and those patients probably had marginal levels in blood and tissues. A cancer cell population evolving in that milieu may have been extraordinarily sensitive to supplemental vitamin C.

In 1975, Cameron et al. presented an extraordinary case of a double complete regression in a 42-year-old male patient diagnosed with reticulum cell sarcoma treated with vitamin C [21]. The patient had been scheduled for radiotherapy and chemotherapy, but due to rapid deterioration and administrative delays, vitamin C was implemented at 10 g/d intravenously for 10 days, followed by 10 g/d taken orally. The response to vitamin C was rapid: the patient felt well within 10 days, elevated erythrocyte sedimentation rates and serum seromucoid levels decreased, and chest x-rays taken 14 and 22 days after initiation of treatment showed resolution of mediastinal and hilar masses. The patient returned to his job but complained of symptoms about 5 months after initial treatment and 4 weeks after ceasing oral vitamin C supplementation. Blood tests and x-rays confirmed the recurrence of cancer, and the patient was given 10 g/d of oral vitamin C for 2 weeks without apparent benefit as the disease progressed. Intravenous vitamin C (20 g/d) was implemented and continued for 14 days, followed by 12.5 g/d of oral vitamin C. Six months after starting the second vitamin C treatment, the patient's chest x-rays were normal, and he had resumed employment.

In 1976, Cameron and Pauling published the results of their large-scale clinical study in the *Proceedings of the National Academy of Science* [22]. This high-profile publication attracted much attention in the media and among physicians. One hundred hospitalized terminal cancer patients treated with vitamin C (usually 10 g/d intravenously for 10 days followed by 10 g/d orally) were matched with 1000 patients not treated with vitamin C in the Vale of Leven Hospital in Scotland. Patients had been previously treated conventionally with radiotherapy, surgery, hormones, or cytotoxic substances. As clarified later, only 4 vitamin C–treated patients had received prior chemotherapy, and only 20 had received prior radiotherapy. Consensus by at least two physicians determined untreatability of the vitamin C–treated patients. Fifty patients in the vitamin C group were described previously by Cameron and Campbell [18], and another 50 were randomly selected from the hospital record index of vitamin C–treated patients. For each vitamin C–treated patient, 10 terminal patients closely matched for age, sex, and cancer site and type were selected randomly from the hospital record index from the preceding

Linus Pauling and Ewan Cameron at the Linus Pauling Institute of Science and Medicine, Menlo Park, California (ca. 1978).

10 years and comprised the control group. Twenty percent of the controls were hospitalized contemporaneously with the vitamin C–treated patients. Most vitamin C–treated patients had cancer of the bronchus (15), colon (13), stomach (13), breast (11), or kidney (9). Patient records of the control group were analyzed by a third party to ascertain date of untreatability so that survival times in the two groups could be calculated and compared. The subsequent analysis revealed that the mean survival time for vitamin C–treated patients was about 4.2 times longer than that of controls (210 days versus 50 days, respectively) and that the rate of death for about 90% of the vitamin C–treated patients was one-third the rate for controls. The remaining 10% of vitamin C–treated patients survived, on average, more than 20 times longer than controls.

Despite these encouraging results, doubts were raised by other investigators concerning the lack of formal randomization, whether the two groups were representative of the same population, and the determination of untreatability. In their next joint paper, Cameron and Pauling addressed these concerns [23]. For the new analysis, a new set of control patients was selected, and 10 vitamin C–treated patients with rare cancers were replaced by 10 other vitamin C–treated patients who could be matched better with controls. The new control group substantially overlapped with the previous control group and included 370 patients who were hospitalized concurrently with the vitamin C–treated patients. New dates were acquired for the vitamin C–treated group and the control group: (1) first appearance of symptoms, (2) first hospitalization, (3) initiation of any therapy deemed to be potentially curative, (4) initiation of palliative treatment, (5) determination of untreatability when therapy was discontinued, and (6) death. Analysis revealed that the time between first hospital attendance and determination of untreatability was not significantly different for the two groups, thus confirming that the two groups were representative subpopulations of the cancer population in the hospital. The mean survival time of the vitamin C–treated patients was about 300 days longer than that of the patients in the control group. Twenty-two percent of the vitamin C–treated patients and 0.4% of patients in the control group survived longer than 1 year after the date of untreatability. At the time of analysis, 8 of the vitamin C–treated patients were still alive, whereas all the patients in the control group had died. The vitamin C–treated patients in this study had a variety of cancers, including cancers of the colon (17), bronchus (17), stomach (13), and breast (11). Patients in the vitamin C–treated group with cancers of the colon, breast, bladder, or kidney had mean survival times of over 300 days longer from the date of untreatability than matched controls. Vitamin C–treated patients with cancers of the rectum had mean survival times of over 200 days longer from the date of untreatability than matched controls.

Cameron and Pauling had become convinced that vitamin C provided substantial benefits to cancer patients and wanted the NCI to fund an RCT that would provide definitive results. Cameron was convinced that vitamin C was valuable in treating cancer and was therefore unwilling to conduct such a placebo-controlled trial himself because he thought it would be unethical for him to withhold vitamin C from his patients. He and Pauling were also frustrated that five grant requests submitted to the NCI from 1973 to 1977 to study vitamin C and cancer had been rejected [24].

EARLY CLINICAL STUDIES IN JAPAN

Pauling was also aware of clinical work on vitamin C and cancer conducted in Japan by Morishige and Murata that had not garnered much attention. In 1955, Morishige became interested in the inhibition by vitamin C of posttransfusional viral hepatitis in surgical patients. In a somewhat indeterminate period ending in 1973, 53 cancer patients admitted to the Fukuoka Torikai Hospital in Fukuoka, Japan, received no supplemental vitamin C, 90 received 0.5 g/d, and 18 received 1 g/d. By 1977, all cancer patients admitted to the hospital received at least 5 g/d of vitamin C. In 1979, Morishige and Murata published the results of their 5-year study (1973–1977) in which patients with terminal cancer of various types (stomach, $n = 34$; lung, $n = 15$; uterus, $n = 12$) were given daily doses of vitamin C ranging from 4 g or less to as much as 60 g [25]. Patients were diagnosed as terminal about 10 days after hospital admission. This study was not designed as an RCT. Instead, 44 patients with advanced cancer were given vitamin C at low doses (≤ 4 g/d, mean = 1.5 g/d), and 55 were given high doses (≥ 5 g/d, mean = 29 g/d). The sex and age distributions for patients in both groups were similar. The mean survival time for patients on the low-dose vitamin C regimen was 43 days; mean survival time for patients on the high-dose vitamin C regimen was 201 days. By mid-1978, 6 patients receiving high-dose vitamin C were still alive, representing 11% of patients in that group, with an average survival time of 866 days after being deemed terminal. These results were similar to those reported by Cameron and Pauling in 1976. All patients in the low-dose vitamin C group had died by day 174. The investigators concluded that vitamin C was of value in significantly extending survival time of terminal cancer patients, especially in cases of uterine cancer, and that their study substantiated the work of Cameron and Pauling.

Morishige and Murata did not specify the route of vitamin C administration in the 1979 report, but a subsequent paper in 1982 [26] on two sets of cancer patients treated with vitamin C, one in the Fukuoka hospital previously reported and one in the Kamioka Kozan Hospital in Kamioka, Japan, noted that high-dose vitamin C was provided orally (usually 6–12 g/d) and intravenously (10 or 12 g per 500 mL). In the second paper, the authors reported that, as of April 1980, 3 patients in the high-vitamin C group were still alive, with survival times of more than 1561 days, more than 1419 days, and more than 1671 days. The patients were described as "clinically well," with neither progression nor regression of their tumors, thus representing tumor stasis. Further analysis of the survival times for patients in the low- and high-dose vitamin C groups revealed that the average survival time of patients after determination of terminal cancer in the low-dose vitamin C group was 43 days and that of patients in the high-dose vitamin C group was more than 246 days, or about 5.7 times longer.

The design of the Kamioka Hospital trial was similar to that of the Fukuoka trial: patients with terminal cancer were given low-dose vitamin C or high-dose vitamin C. Additionally, a control group without supplemental vitamin C was selected. Beginning in 1976, 6 patients received low-dose vitamin C (0.5–3 g/d). In 1977, 6 other patients began receiving high-dose vitamin C (5–30 g/d). Nineteen patients did not receive any supplemental vitamin C and served as the control group. The trial continued until the end of 1979. By April 1980, none of the control patients had survived longer than 98 days, with an average survival time of 48 days, whereas the average survival times for patients in the low- and high-dose vitamin C groups were 84 days and 115 days, respectively. One patient in the high-dose vitamin C group was still alive in April 1980 and exhibited a decrease in bladder tumor volume from 115 to 7. The authors concluded that the administration of vitamin C to terminal cancer patients increased survival times, decreased reliance on narcotics to ameliorate pain, and improved the quality of life.

MAYO CLINIC CLINICAL TRIALS

The National Institutes of Health (NIH) funded an RCT on vitamin C and cancer at the Mayo Clinic, which was published in 1979 [27]. Patients with untreatable cancer (about half of them with colorectal or pancreatic cancer) were randomized into two groups with similar age, sex, type of cancer, and performance score (Eastern Cooperative Oncology Group scale) characteristics. The vitamin C–treated patients ($n = 60$) got 10 g/d of vitamin C only orally in 20, 500-mg capsules in four divided doses; the control group ($n = 63$) received 20 capsules of a lactose placebo in an identical schedule. The treatment continued until death or until the patient was unable to take the capsules by mouth. The median survival time for all patients in the trial was around 7 weeks. Fewer than 10% of patients in either group had not received prior therapy. About 28% had received prior radiotherapy, and about 88% had received prior chemotherapy. Data analysis demonstrated that there was no significant difference in survival rates between the treatment and placebo groups (log-rank test, $p = 0.61$). Symptoms and side effects were also about the same among patients in either group, although a higher percentage of vitamin C–treated patients reported increased strength compared to patients in the control group (26% versus 13%, respectively) and better pain control (24% versus 15%, respectively). The authors discussed the possibility that prior therapy had impaired immunocompetence of patients in their study and concluded that supplemental vitamin C had no benefit for patients with advanced cancer who had received prior radiotherapy or chemotherapy.

Pauling had been largely responsible for persuading Vincent DeVita of the NCI to fund the RCT of vitamin C and cancer. In communications with Charles Moertel of the Mayo Clinic, who was selected to conduct the RCT, Cameron and Pauling emphasized the principle of host resistance, which required immunocompetence. Cameron noted that he had little experience with cytotoxic chemotherapy because it was not standard practice in his hospital unit. Furthermore, the authors of the Mayo Clinic report had erroneously stated that 50% of vitamin C–treated patients in Cameron and Pauling's first study had been previously treated with radiotherapy and chemotherapy, whereas only 4% of the vitamin C–treated patients had received such prior therapies [11,28]. Cameron and Pauling also complained that the vitamin C supplementation in the Mayo Clinic trial had been stopped abruptly, which, they explained, could lead to a rebound effect (rapid depletion of vitamin C in tissues to abnormally low levels after the body had become conditioned to much higher intakes, possibly affecting cancer progression). While communication between the two teams of investigators had been initially cordial, as reports in the media about the failure of vitamin C proliferated, the communications became more acrimonious and recriminatory. Cameron and Pauling were upset that the Mayo Clinic study's methodology had been falsely represented as replicating Cameron's. One issue noted by Pauling that was not emphasized at the time, perhaps for reasons alluded to previously, was the mode of vitamin C's administration: whereas Cameron had initiated patients with 10 g/d of vitamin C intravenously, followed by oral administration, the Mayo Clinic investigators provided only oral vitamin C to patients [11].

In 1985, the Mayo Clinic team published a second RCT of vitamin C and cancer that attempted to respond to criticisms about their earlier RCT [29]. In this second trial, 100 male and female patients with advanced colorectal cancer who had no prior chemotherapy were randomly assigned to a treatment group ($n = 51$) given 10 g/d of vitamin C (a protocol identical to their previous study) or to a control group ($n = 49$). Patients in both groups had comparable characteristics and were ambulatory. As in their previous study, patients in the treatment group received vitamin C only orally, not intravenously. Treatment with vitamin C or placebo continued until the patient was unable to tolerate oral medication or until cancer progressed, as defined by a 50% increase in the product of perpendicular dimensions of the tumor, new appearance of a malignancy, substantial worsening of symptoms or performance status, or a decrease of at least 10% in body weight. The median duration of treatment in the vitamin C group was 2.5 months (range, 1 day-15.6 months) and that of the control group was 3.6 months (range, 7 days-25.5 months). Compliance was ascertained by the patients' recording of drug intake. Urine samples were also obtained from 11 participants during the study. Five were in the vitamin C–treatment group and had 24-hour urinary vitamin C values of 2 g or more; five patients in the control group had levels of 0.55 g or less, while one patient in the control group had an intermediate urinary vitamin C value. The authors ascribed this anomaly to various drugs taken by the subject that may have interfered with the nonspecific vitamin C assay.

Analysis of the data showed that the median time from initiation of therapy to disease progression was 2.9 months for vitamin C–treated patients and 4.1 months for patients in the control group. In a somewhat ambiguous statement, the authors noted that half of the 58 patients "who have discontinued participation in this study" then received fluorouracil and/or other therapies. At the time of publication, 85 of the study patients had died. The survival curves for vitamin C–treated patients and controls were about the same for the first year (49% versus 47%, respectively). The difference between survival curves increased 12 months after the onset of treatment, and by 24 months, all of the vitamin C–treated patients had died, whereas about 10% of the control patients were still alive after 39 months. The authors concluded that high-dose vitamin C has no value against advanced malignant cancer and that Cameron's positive results must have been due to case-selection bias.

Cameron and Pauling criticized the second Mayo Clinic trial on several grounds: (1) based on the protocol compliance comments, it appeared that at least several patients in the control group had larger than expected urinary excretion of vitamin C, indicating that they may have surreptitiously taken vitamin C supplements; (2) vitamin C treatment was not continued indefinitely—it ceased after a median duration of 2.5 months and, in some cases, was replaced with cytotoxic therapies; and (3) the termination of vitamin C treatment may have induced a rebound effect [16]. Very contentious communications followed between Pauling and Arnold Relman, the editor of the *New England Journal of Medicine* in which the two Mayo Clinic trials were published, and Creagan and Moertel, who carried out the Mayo Clinic trials. Pauling was also critical of the guest editorial in the *New England Journal of Medicine* in 1985 by Robert Wittes of the NCI, who described the second Mayo Clinic trial as "clear-cut" and "methodologically sound and therefore definitive." Wittes stated that "Ascorbic acid was given in the same daily dose and by the same route advocated by Cameron and Pauling" [30].

Largely overlooked in the controversy was the mode of administration of vitamin C, which was erroneously described by Wittes as identical in the Mayo Clinic trials to Cameron's work in Scotland. Of course, the Mayo Clinic researchers gave vitamin C only orally; Cameron's patients were typically commenced on intravenous vitamin C and continued indefinitely with oral vitamin C. At the time, the vast difference in resultant concentrations of vitamin C in blood from oral versus intravenous vitamin administration was not amply appreciated, although Cameron and Pauling wrote in 1979, "The fraction of a large oral dose [of vitamin C] that is transferred across the intestinal wall into the blood stream may be less than 50%. Accordingly, intravenous ascorbate may be more than twice as effective as oral ascorbate" [11]. They also noted that a dose of 10 g of vitamin C maximizes the concentration of vitamin C in white blood cells in healthy subjects, while comparable vitamin C intake raised the level in white blood cells from Cameron's cancer patients but never achieved saturation. Compromised leukocyte function might affect immunocompetence, a central tenet of Cameron's and Pauling's postulated anticancer mechanisms of vitamin C.

In 1989, Pauling and Herman published a detailed analysis of the Mayo Clinic results from a statistical perspective [31]. Using the Hardin Jones biostatistical principle that the death rate of cancer patients in a homogeneous cohort is constant, they presented three criteria for the validity of cancer trials: (1) "the treatment of all members of the cohort should be the same, and it should be continuous and unchanged from the time $t = 0$ when the patient enters the trial until the time t when the patient dies or t+ when, without dying, is withdrawn from the set of survivors at risk"; (2) there should not be any significant lag period from $t = 0$ without observed deaths ("the Hardin Jones straight line passes through the 100% axis at $t = 0$"); and (3) in a heterogeneous cohort representing subcohorts with different life expectancies, the "semi-logarithmic survival curve must bend away from the Hardin Jones initial straight line only in the direction of increased survival times for the longer-term survivors." After reviewing the results of other clinical trials of cancer patients, Pauling and Herman found that most met the criteria and were valid but that one failed to meet all three criteria: the second Mayo Clinic trial.

CLINICAL STUDIES BY HOFFER AND PAULING

While Cameron, Pauling, and other clinical researchers, including Hoffer, who was a Canadian research psychiatrist, and Riordan in Kansas [32], continued their work with vitamin C and cancer, the conventional medical community lost interest in vitamin C, having become convinced by the Mayo Clinic trials that vitamin C was worthless against cancer. Cameron died in 1991, and Pauling continued to work with Hoffer on analyses of the response of patients with advanced cancer to an oral daily regimen that included 12 g of vitamin C (range, 3–40 g), high-dose B vitamins (1.5 or 3 g of niacin or niacinamide, 250 mg of pyridoxine, and others), vitamin E (800 IU), β-carotene (30,000 IU), selenium (200–500 μg), and other minerals, as well as dietary advice [33]. In Pauling's opinion, this expanded regimen seemed to be more effective than vitamin C alone. Patients with advanced, untreatable cancer had been referred to Hoffer for management of depression and anxiety. Hoffer advised taking the oral supplement regimen in addition to appropriate conventional therapy and observed that patients lived much longer than expected. The study involved 134 such ambulatory patients, about 85% of whom had undergone conventional therapy, including surgery, radiation, or chemotherapy; the control group was composed of 33 patients who elected not to follow the regimen but otherwise were not substantially different from patients in the treatment group. Analysis of treatment cohorts (one consisting of 40 females with cancer of the breast, ovary, uterus, or cervix; the other representing 61 patients with other cancers) revealed that about 80% of patients in the first cohort survived 21 times longer than controls and that 80% of patients in the second cohort survived 13 times longer than controls. The mean survival time for controls was 5.7 months, whereas the mean survival time for 80% of patients who followed the regimen was 92 months. Pauling and

Abram Hoffer, Rose Hoffer, and Linus Pauling in San Francisco (1991).

Hoffer also suggested that physicians give high-dose vitamin C by intravenous infusion to patients with advanced cancer.

A second report on patients in the first study followed for an additional 31 months found that 40% of patients who followed the regimen survived 5 or more years and that 60% had survival times about four times longer than controls [34]. Results for an additional 170 patients with advanced cancer (155 treated with the regimen and 15 controls) were similar to those reported in the first study. The mean survival time for controls in both the Hoffer studies was 135 days. About 50% of the patients in the extended first study with cancer of the breast, ovary, uterus, or cervix who followed the regimen were categorized as excellent responders (mean survival time greater than 5 years), and about 50% were good responders (mean survival time of 630 days). About 33% of patients with other cancers were excellent responders, and 67% were good responders. Hoffer and Pauling concluded that about 40% of patients with advanced cancer who followed the regimen were excellent responders, whereas about 10% of patients treated only with vitamin C in Cameron's studies were categorized as excellent responders. Since the therapeutic intervention was multifaceted, the relative importance of the various components cannot be ranked.

CYTOTOXICITY AND MODE OF ADMINISTRATION

As discussed, Pauling and Cameron were primarily interested in strengthening host resistance to cancer, which would not necessarily be curative but rather transform cancer into a chronic and manageable disease. Yet Pauling also became aware of the potential cytotoxicity of vitamin C as an anticancer mechanism. In a 1983 paper, Pauling and his coauthors suggested that vitamin C combined with a copper-containing peptide (Cu:GGH; copper:glycylglycylhistidine) designed to mimic the copper transport site of albumin was cytotoxic to mouse Ehrlich ascites tumor cells *in vitro* due to the generation of hydrogen peroxide; the cytotoxicity was nullified by the addition of catalase to the culture medium [35]. The generation of cytotoxic hydrogen peroxide by high extracellular concentrations of vitamin C achieved only by intravenous infusion has been documented in detail by Levine's group at the NIH [20] and is currently recognized as a major anticancer mechanism of vitamin C. Variable clinical results may be explained partly by the presence of catalase in some cancer cells, rendering them less susceptible to hydrogen peroxide [36], although other studies have not found an association between catalase activity in cancer cell lines and vitamin C–mediated toxicity [20].

Morishige of the Nakamura Memorial Hospital in Fukuoka, Japan, developed a clinical protocol in the early 1980s involving the administration of Cu:GGH with intravenous high-dose vitamin C [37]. Morishige postulated that the high peptide-cleaving activity of tumor cells would decompose the Cu:GGH compound into reactive copper, which, when exposed to vitamin C, would generate free radicals lethal to tumor cells. Copper bound to GGH would remain relatively unreactive and, therefore, would not be expected to damage normal cells or produce undesirable side effects. Morishige had found that either Cu:GGH or low concentrations of vitamin C did not exhibit significant cytotoxicity against Ehrlich ascites tumor cells *in vitro*. However, the intraperitoneal injection of vitamin C and Cu:GGH into mice inoculated with Ehrlich ascites tumor cells significantly increased survival times compared to controls or to mice injected with only Cu:GGH or vitamin C. Based on these observations, Morishige treated a 34-year-old woman admitted to the hospital with severe pain due to an inoperable osteosarcoma with intra-arterial injections of Cu:GGH and vitamin C for 2 months. Treatment alleviated pain, and 4 months after the commencement of therapy, x-rays revealed complete regression and calcification of the lesion.

Levine's pharmacokinetic research [38] established that there are vast differences in the concentration of vitamin C in blood after oral or intravenous administration. In a rigorously designed study with seven healthy, young men, Levine found that average plasma concentrations of vitamin C after oral doses of 100, 400, or 2500 mg ($n = 3$) were 56, 70, and 85 μM, respectively; plasma concentrations did not exceed 100 μM. However, plasma concentrations after intravenous administration of 1.25 g of vitamin C approached millimole levels but returned to baseline after about 7 hours. In a subsequent publication [39], Padayatty and Levine noted that the intravenous administration of 5–10 g of vitamin C, which is the range typically used in Cameron's early clinical work, may produce plasma concentrations of about 5 mM. In 2005, Chen et al. [20] reported that the cytotoxicity of vitamin C in cancer cell lines was critically dependent on the formation of hydrogen peroxide, which increased linearly in the culture media with the formation of the ascorbyl radical. Normal cells were unaffected by vitamin C concentrations as high as 20 mM, whereas cancer cells were sensitive to vitamin C concentrations less than 4 mM, which are achievable by intravenous administration.

The influence of vitamin C on hypoxia-inducible factor (HIF) has also been investigated. HIF, largely quiescent in normal cells but active in tumors, promotes the growth of the vasculature and contributes to tumor growth. One of its two components, HIF-1α, is significantly decreased by vitamin C. Comparing tumor and normal tissue from 49 colorectal cancer patients, Kuiper et al. [40] found an inverse relationship between vitamin C content in tumors and HIF activation and tumor size. Higher vitamin C content in tumors also correlated with improved disease-free survival 6 years after surgery.

Host resistance as proposed by Cameron and Pauling has not attracted much research attention, and it remains speculative. Under Pauling's direction, much research at the Linus Pauling Institute of Science and Medicine in the 1970s addressed the preventative anticancer effect of vitamin C on hairless mice exposed to ultraviolet radiation [41] and, in the early 1990s, on the structural characteristics of derivatives and oxidation products of vitamin C that accounted for cytotoxicity in cell culture and animal models. In particular, the researchers found that (1) the enediol lactone ring of several ascorbate isomers and ascorbyl palmitate was associated with substantial cytotoxicity in a murine leukemia cell line [42] and (2) the cytotoxicity of vitamin C was increased significantly in the presence of copper and iron ions [43,44].

FUTURE DIRECTIONS

Since Pauling's death in 1994, a number of clinical trials and anecdotal case reports using intravenous vitamin C have been published (for reviews and references, see [45] and [46]). Many of these have used vitamin C in conjunction with chemotherapeutic drugs like arsenic trioxide, dexamethasone, bortezomib, melphalan, or gemcitabine, while other trials have used vitamin C without concomitant chemotherapy or radiation. Generally, the results have been positive, although optimum therapeutic doses, their frequency and duration, and the combination of vitamin C with specific chemotherapeutic drugs or other interventions are critical issues that need to be further explored.

As presented in subsequent chapters of this book, basic research continues to elucidate and amplify the molecular anticancer mechanisms of vitamin C, and this knowledge will inform the design of new clinical studies.

REFERENCES

1. Pauling, L. 1968 Orthomolecular psychiatry. *Science* 160, 265–271.
2. Ames, B. N., Elson-Schwab, I. and Silver, E. A. 2002 High-dose vitamin therapy stimulates variant enzymes with decreased coenzyme binding affinity (increased K(m)): Relevance to genetic disease and polymorphisms. *Am. J. Clin. Nutr.* 75, 616–658.
3. Pauling, L. 1974 Some aspects of orthomolecular medicine. *J. Internatl. Acad. Prev. Med.* 1, 1–30.
4. Pauling, L. 1986 *How to Live Longer and Feel Better.* W. H. Freeman, New York.
5. Pauling, L. 1970 *Vitamin C and the Common Cold.* W. H. Freeman, San Francisco.
6. Hemila, H. 1994 Does vitamin C alleviate the symptoms of the common cold? A review of current evidence. *Scand. J. Infect. Dis.* 26, 1–6.
7. Pauling, L. 1970 Evolution and the need for ascorbic acid. *Proc. Natl. Acad. Sci. USA* 67, 1643–1648.
8. Cameron, E. 1966 *Hyaluronidase and Cancer.* Pergamon Press, New York.
9. McCormick, W. J. 1954 Cancer: The preconditioning factor in pathogenesis. *Arch. Pediatr.* 71, 313–322.
10. McCormick, W. J. 1959 Cancer: A collagen disease, secondary to a nutritional disease? *Arch. Pediatr.* 76, 166–171.
11. Cameron, E. and Pauling, L. 1979 *Cancer and Vitamin C.* Linus Pauling Institute of Science and Medicine, Palo Alto.
12. Deucher, W. G. 1940 Observations on the metabolism of vitamin C in cancer patients (in German). *Strahlentherapie* 67, 143–151.
13. Cameron, E. and Pauling, L. 1979 Ascorbic acid as a therapeutic agent in cancer. *J. Internatl. Acad. Prevent. Med.* 5, 8–29.
14. Greer, E. 1954 Alcoholic cirrhosis complicated by polycythemia vera and then myelogenous leukemia and tolerance of large doses of vitamin C. *Med. Times* 82, 865–868.
15. Cheraskin, E., Ringsdorf Jr., W. M., Hutchins, K., Setyaadmadja, A. T. S. H. and Wideman, G. L. 1968 Effect of diet upon radiation response in cervical carcinoma of the uterus: A preliminary report. *Acta Cytologica* 12, 433–438.
16. Richards, E. 1991 *Vitamin C and Cancer: Medicine or Politics?* Macmillan, London.
17. Young, C. 1976 Memorandum of April 8. Special Collections & Archives Research Center, Oregon State University.
18. Cameron, E. and Campbell, A. 1974 The orthomolecular treatment of cancer II. Clinical trial of high-dose ascorbic acid supplements in advanced human cancer. *Chem.-Biol. Interact.* 9, 285–315.
19. Tsukaguchi, H., Tokui, T., Mackenzie, B., Berger, U. V., Chen, X. Z., Wang, Y., Brubaker, R. F. and Hediger, M. A. 1999 A family of mammalian Na+-dependent L-ascorbic acid transporters. *Nature* 399, 70–75.
20. Chen, Q., Espey, M. G., Krishna, M. C., Mitchell, J. B., Corpe, C. P., Buettner, G. R., Shacter, E. and Levine, M. 2005 Pharmacologic ascorbic acid concentrations selectively kill cancer cells: Action as a pro-drug to deliver hydrogen peroxide to tissues. *Proc. Natl. Acad. Sci. USA* 102, 13604–13609.
21. Cameron, E., Campbell, A. and Jack, T. 1975 The orthomolecular treatment of cancer III. Reticulum cell sarcoma: Double complete regression induced by high-dose ascorbic acid therapy. *Chem.-Biol. Interact.* 11, 387–393.
22. Cameron, E. and Pauling, L. 1976 Supplemental ascorbate in the supportive treatment of cancer: Prolongation of survival times in terminal human cancer. *Proc. Natl. Acad. Sci. USA* 73, 3685–3689.
23. Cameron, E. and Pauling, L. 1978 Supplemental ascorbate in the supportive treatment of cancer: Reevaluation of prolongation of survival times in terminal human cancer. *Proc. Natl. Acad. Sci. USA* 75, 4538–4542.

24. Cameron, E. and Pauling, L. 1978 Experimental studies designed to evaluate the management of patients with incurable cancer. *Proc. Natl. Acad. Sci. USA* 75, 6252.
25. Morishige, F. and Murata, A. 1979 Prolongation of survival times in terminal human cancer by administration of supplemental ascorbate. *J. Interntl. Acad. Prev. Med.* 5, 47–52.
26. Murata, A., Morishige, F. and Yamaguchi, H. 1982 Prolongation of survival times of terminal cancer patients by administration of large doses of ascorbate. *Interntl. J. Vitam. Nutr. Res. Suppl.* 23, 103–113.
27. Creagan, E. T., Moertel, C. G., O'Fallon, J. R., Schutt, A. J., O'Connell, M. J., Rubin, J. and Frytak, S. 1979 Failure of high-dose vitamin C (ascorbic acid) therapy to benefit patients with advanced cancer. *N. Engl. J. Med.* 301, 687–690.
28. Pauling, L. 1980 Vitamin C therapy of advanced cancer. *N. Engl. J. Med.* 302, 694.
29. Moertel, C. G., Fleming, T. R., Creagan, E. T., Rubin, J., O'Connell, M. J. and Ames, M. M. 1985 High-dose vitamin C versus placebo in the treatment of patients with advanced cancer who have had no prior chemotherapy: A randomized double-blind comparison. *N. Engl. J. Med.* 312, 137–141.
30. Wittes, R. E. 1985 Vitamin C and cancer. *N. Engl. J. Med.* 312, 178–179.
31. Pauling, L. and Herman, Z. S. 1989 Criteria for the validity of clinical trials of treatments of cohorts of cancer patients based on the Hardin Jones principle. *Proc. Natl. Acad. Sci. USA* 86, 6835–6837.
32. Riordan, H. D., Jackson, J. A. and Schultz, M. 1990 Case study: High-dose intravenous vitamin C in the treatment of a patient with adenocarcinoma of the kidney. *J. Orthomol. Med.* 5, 5–7.
33. Hoffer, A. and Pauling, L. 1990 Hardin Jones biostatistical analysis of mortality data for cohorts of cancer patients with a large fraction surviving at the termination of the study and a comparison of survival times of cancer patients receiving large regular oral doses of vitamin C and other nutrients with similar patients not receiving those doses. *J. Orthomol. Med.* 5, 143–154.
34. Hoffer, A. and Pauling, L. 1993 Hardin Jones biostatistical analysis of mortality data for a second set of cohorts of cancer patients with a large fraction surviving at the termination of the study and a comparison of survival times of cancer patients receiving large regular oral doses of vitamin C and other nutrients with similar patients not receiving those doses. *J. Orthomol. Med.* 8, 157–167.
35. Kimoto, E., Tanaka, H., Gyotoku, J., Morishige, F. and Pauling, L. 1983 Enhancement of antitumor activity of ascorbate against Ehrlich ascites tumor cells by the copper: Glycylglycylhistidine complex. *Cancer Res.* 43, 824–828.
36. Klingelhoeffer, C., Kammerer, U., Koospal, M., Muhling, B., Schneider, M., Kapp, M., Kubler, A., Germer, C.-T. and Otto, C. 2012 Natural resistance to ascorbic acid induced oxidative stress is mainly mediated by catalase activity in human cancer cells and catalase-silencing sensitizes to oxidative stress. *BMC Complement. Altern. Med.* 12, doi: 10.1186/1472-6882-12-61
37. Morishige, F. ca. 1983 Studies on the role of large dosage of vitamin C in the orthomolecular nutritional treatments of cancer. Self published.
38. Levine, M., Conry-Cantilena, C., Wang, Y., Welch, R. W., Washko, P. W., Dhariwal, K. R., Park, J. B. et al. 1996 Vitamin C pharmacokinetics in healthy volunteers: Evidence for a recommended dietary allowance. *Proc. Natl. Acad. Sci. USA* 93, 3704–3709.
39. Padayatty, S. J. and Levine, M. 2001 New insights into the physiology and pharmacology of vitamin C. *Can. Med. Assoc. J.* 164, 353–355.
40. Kuiper, C., Dachs, G. U., Munn, D., Currie, M. J., Robinson, B. A., Pearson, J. F. and Vissers, M. C. M. 2014 Increased tumor ascorbate is associated with extended disease-free survival and decreased hypoxia-inducible factor-1 activation in human colorectal cancer. *Front. Oncol.* 4,10, doi: 10.3389/fonc.2014.00010.
41. Pauling, L., Willoughby, R., Reynolds, R., Blaisdell, B. E. and Lawson, S. 1982 Incidence of squamous cell carcinoma in hairless mice irradiated with ultraviolet light in relation to intake of ascorbic acid (vitamin C) and of D, L-alpha-tocopherol acetate (vitamin E). *Int. J. Vitam. Nutr. Res. Suppl.* 23, 53–82.
42. Leung, P. Y., Miyashita, K., Young, M. and Tsao, C. S. 1993 Cytotoxic effect of ascorbate and its derivatives on cultured malignant and nonmalignant cell lines. *Anticancer Res.* 13, 475–480.
43. Leung, P. Y., Dunham, W. B. and Tsao, C. S. 1992 Ascorbic acid with cupric ions as a chemotherapy for human lung tumor xenografts implanted beneath the renal capsule of immunocompetent mice. *In Vivo* 6, 33–40.

44. Tsao, C. S., Dunham, W. B. and Leung, P. Y. 1995 Growth control of human colon tumor xenografts by ascorbic acid, copper, and iron. *Cancer J.* 8, 157–163.
45. Cameron, E. and Pauling, L. 2018 *Cancer and Vitamin C: The 21st-Century Edition*. Camino Books, Philadelphia.
46. Nauman, G., Gray, J. C., Parkinson, R., Levine, M. and Paller, C. J. 2018 Systematic review of intravenous ascorbate in cancer clinical trials. *Antioxidants* 7, 89, doi: 10.3390/antiox7070089
47. Ava Helen and Linus Pauling Papers 1960 OSU Libraries Special Collections & Archives Research Center.

PART II

Mechanisms of Action for Vitamin C in Cancer

CHAPTER TWO

High-Dose Intravenous Ascorbate as a Pro-Oxidant Inducing Cancer Cell Death

Ping Chen and Qi Chen

CONTENTS

INTRODUCTION

From a well-described account by Stephen Lawson in this book (Chapter 1), one can see that the role of ascorbate as a treatment for cancer has been quite controversial in the past few decades. In the 1970s, observational reports by Pauling and Cameron described that ascorbate, given at high doses (pharmacologic doses) of 10 g daily, was effective in treating some cancers and in improving survival and well-being in terminal cancer patients [1–4]. Pauling's reports generated much excitement as well as criticism among cancer scientists, oncologists, physicians, and the general public. Whereas the general public tended to embrace these findings, the criticisms mainly came from the scientific and medical communities. The scientific issues raised included that these studies were retrospectively controlled case series; therefore, they lacked randomization and blinding. There were possibilities that the positive results were due to a placebo effect and lack of a proper control group. The patient population was in rural Scotland, and they had vitamin C deficiency to begin with; therefore, the results could not apply to a more generalized patient population [5–7]. Subsequently, two formal clinical trials with randomization, prospective placebo control, and double blinding were conducted at the Mayo Clinic, led by Charles Moertel, using 10 g ascorbate daily. The first trial was in terminal cancer patients, and the second trial was in newly diagnosed cancer patients without pretreatment. Unfortunately, both trials showed no effect of ascorbate [5,6]. Ascorbate seemed to have no role in cancer therapy and was discarded as a cancer treatment by the conventional medical community.

The use of high-dose ascorbate, however, continued, both orally and parenterally. In fact, vitamin C is probably the most popular single vitamin consumed by the general public in the United States, with sales of $1.2 billion in the United States in 2017 according to the *Nutrition Business Journal* [8]. The use is commonly for improving general health or for optimal health. Oral vitamin C is widely used to prevent or treat the common cold, which was also promoted by Linus Pauling [9]. For treatment of cancer and a number of other diseases, large doses of ascorbate continued to be used, outside the mainstream and in the integrative, complementary and alternative medicine (CAM) communities. To understand the scale of the CAM usage of ascorbate, Padayatty and colleagues delivered survey questions to the CAM practitioners in the United States attending an annual conference in 2006 and 2008, asking about the usage of intravenous ascorbate (IVC), the reasons for using it, and the side effects, if any. When the data came back, it revealed an unexpectedly widespread use [10]. There were at least 8000–10,000 patients treated with ascorbate injections each year just by the 199 survey responders, with an average dose of 28 g every 4 days, and 22 total treatments per patient. Counting this usage in 25 g/50 mL vials of ascorbate, the estimated doses administered to patients were more than 300,000 each year. This number is likely an underestimation of the actual usage, because manufacturers' yearly sales of vials of injectable ascorbate were more than double this number. The top three most common reasons for receiving the high-dose ascorbate treatment were infection, cancer, and fatigue. The survey found that side effects were minor—less than 1% of patients had side effects, and most of them were minor to negligible, which included lethargy/fatigue, change in mental status (all such cases listed by a single practitioner and without details), vein irritation, and nausea and vomiting.

Despite the widespread use and apparent safety, there seemed to be a lack of scientific rationale, at least in the case of treating cancers, after the Mayo Clinic trials in the 1980s determined that ascorbate had "no utility in cancer treatment" [11]. But as Robert Wittes, the editor of the *New England Journal of Medicine* at the time, disclaimed in his editorial with the second Mayo Clinic trial: this conclusion might change, if new evidence arose [11].

NEW EVIDENCE FOR REEXAMINING THE ROLE OF ASCORBATE IN CANCER TREATMENT

New evidence first arose from a rather unexpected angle. At the National Institute of Diabetes and Digestive and Kidney Diseases (NIDDK), Mark Levine and colleagues wanted to quantitatively define optimal nutrition and strategies that could be translated for use by the general public [12,13]. They selected vitamin C as a model nutrient for water-soluble vitamins and conducted a pharmacokinetic study to detail the dose-concentration relationship in humans following oral ingestion [14,15]. Their study indicated that multiple mechanisms controlled the physiologic homeostasis of vitamin C in humans. These mechanisms include intestinal absorption, tissue/cellular uptake/accumulation, renal reabsorption and excretion, and rate of utilization. The sodium-dependent vitamin C transporters (SVCTs) are responsible for intestinal absorption, renal reabsorption, and cellular uptake of ascorbate. In healthy men and women, once oral intake exceeded 250 mg daily, the plasma concentration of vitamin C reached a plateau at around 80 μM. Higher oral doses would not increase plasma and tissue concentrations much more. Due to saturations in the capacity of the SVCTs, the intestinal absorption saturates, and the renal reabsorption saturates, and excretion increases. Bioavailability of oral vitamin C decreases when dose increases. As a result, a tightly controlled equilibrium is maintained in human plasma and cells for oral ingestion of ascorbate.

The tight control in bioavailability can be bypassed when ascorbate is administered intravenously. Using 1.25 g IVC, the National Institutes of Health (NIH) investigators achieved plasma concentrations of 1200 μM and found that the millimolar range concentrations were maintained for a few hours in the bloodstream until renal excretion restored the equilibrium. This high concentration was never reached by oral ingestion. The same oral dose provided plasma concentrations about 100 μM, less than 1/10 of that of intravenous injection [14,16]. It is expected that the difference between the plasma concentrations that oral and IVC can provide will be even bigger as the dose further increases. With these data, Levine had a surprising realization: Pharmacokinetics of ascorbate had been overlooked in the earlier cancer studies [7]. While both the Mayo Clinic trials and Pauling's studies used the same dose of 10 g/day, the routes of administration were different. The Pauling and Cameron patients received both oral and intravenous vitamin C, whereas the Mayo Clinic patients received only oral vitamin C. If the 10 g/d is given by continuous IV infusion, as many of Pauling and Cameron's patients received it, it would raise plasma ascorbate concentrations to millimolar range and sustain as long as the infusion continuous. If 10 g of ascorbate is given by a 1–2 hours intravenous infusion, it would produce concentrations more than 5 mM (Figure 2.1). These millimolar concentrations of ascorbate achieved by IV infusion could not be achieved by oral administration. In other words, Pauling and Cameron's patients were exposed to probably more than 25-fold higher blood ascorbate concentrations than the patients in the Mayo Clinic trials.

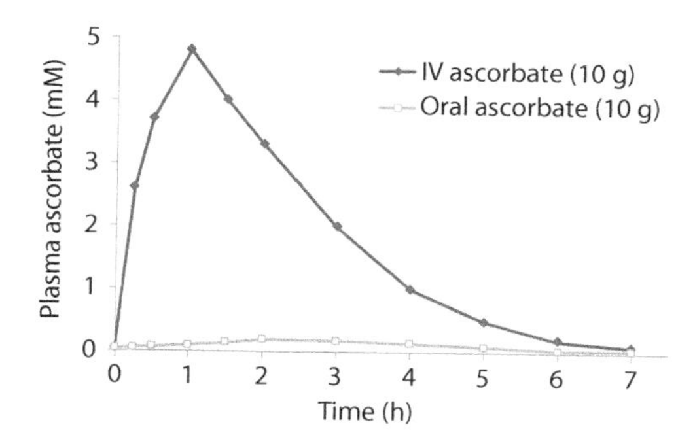

Figure 2.1. Intravenous and oral high-dose ascorbate have different pharmacokinetics. Exemplified here is the dose of 10 g. The difference between oral and IV ascorbate could be bigger as the dose further increases. (From Chen Q. et al. 2015 *Can. J. Physiol. Pharmacol.* 93, 1–9. Reproduced with permission from Canadian Science Publishing.)

These findings provide the first rationale to reexamine the therapeutic role of ascorbate in cancer. Unlike oral ascorbate, IVC, at large doses, could produce plasma concentrations 100-fold higher than maximally tolerated oral doses [7,17]. If IVC was a drug but oral ascorbate was not, then the conclusion was based on a false premise that ascorbate was not effective in cancer treatment. Given the widespread off-label use, the apparent safety, the possible effectiveness of ascorbate, and the lack of efficient treatment to many cancers, the benefit of ascorbate as a cancer therapy is worth reexamining.

ASCORBATE-INDUCED CYTOTOXICITY IS DEPENDENT ON H_2O_2 FORMATION

If ascorbate is effective in treating cancer, the affordability and the expected low toxicity of this vitamin would make it very advantageous. However, although the pharmacokinetic behavior of IVC mimics that of a drug, it does not tell whether ascorbate has the effectiveness of killing a cancer cell. It was essential to first learn whether these supraphysiological concentrations of ascorbate could kill cancer cells, and if so, by what mechanism. Qi Chen and colleagues took the first step and treated cultured cancer cells *in vitro* [18,19]. Concentrations and time courses were selected to mimic IVC administration: cells were exposed to pharmacologic concentrations of ascorbate (0.3–20 mM) for 1–2 hours, and then ascorbate was washed out. A physiologic concentration (0.1 mM) was used as control. Cells were examined for viability 24 hours later. In these studies, a total of 48 cancer cell lines and five types of normal cells were tested. Ascorbate induced cell death in 75% (36 out of 48) of the tested cancer cell lines. For more than half of the cancer cell lines, less than 5 mM of ascorbate was needed to cause a 50% decrease in cell survival (IC_{50} values). All the tested normal cells were insensitive to up to 20 mM of ascorbate, that is, human peripheral white blood cells, fibroblasts, and epithelial cells [18,19]. In another experimental setting with similar ascorbate treatment, long-term survival and clonogenic capacity of the cancer cells were determined, and three out of four tested cancer cell lines had greater than 99% inhibition in the growth and formation of colonies [18].

The death in cancer cells was dependent on extracellular ascorbate concentrations but was not related to the intracellular ascorbate concentrations [18]. Cultured cells do not contain ascorbate unless this vitamin is added to the culture media. In contrast, all human cells *in vivo* contain ascorbate if there is not severe scurvy. Using a lymphoma cell line, Chen et al. preloaded the cells with ascorbate by incubation with physiologic concentrations of ascorbate (100 μM) in the culture media and brought the intracellular concentrations similar to that of normal lymphocytes. Then the preloaded and unloaded lymphoma cells were exposed to the same pharmacologic concentrations of ascorbate. Whether or not preloaded, the lymphoma cells responded with the same amount and type of cell death [18]. Also tested was the oxidized form of ascorbate, dehydroascorbic acid (DHA). Unlike the uptake of ascorbate by SVCTs, DHA is transported into cells by glucose transporters and then reduced to ascorbate by glutathione inside the cells [20]. By adding either 0.5–2 mM of ascorbate or 0.5–2 mM of DHA to the culture media, the investigators achieved equal concentrations of ascorbate inside the lymphoma cells but only observed death when ascorbate was present externally. DHA increased intracellular ascorbate but did not induce cancer cell death even at the highest doses used (2 mM) [18].

In other words, ascorbate-induced cell death was not dependent on preexisting status of ascorbate inside the cells. This observation has important physiologic relevance, because cells in the human body are loaded with ascorbate (given the person is not in persistent scurvy), and the intracellular ascorbate concentrations are relatively stable [14,21–23]. High doses of IVC are expected to only drastically change the concentrations outside the cells. Chen's studies also imply that high-dose IVC could impact cancer cells regardless of whether or not they have intracellular deficiency of ascorbate. This does not exclude, however, the possibility that some subpopulation of ascorbate-deficient cancer cells, or cancer cells with abnormally enhanced uptake of ascorbate, could develop super sensitivity to ascorbate treatment.

A critical role of peroxide formation was identified in the ascorbate-induced cancer cell death [18]. In studying the cell death of the lymphoma cell line treated with ascorbate, Chen et al. noticed a pattern suggestive of H_2O_2-mediated cell death. To determine the mechanism, they tested several H_2O_2 scavengers, including the membrane-impermeant enzyme catalase, the membrane-permeant mimetic of catalase tetrakis(4-benzoic acid) meso-substituted manganoporphyrin (MnTBAP), and the thiol-reducing agent tris(2-carboxyethyl) phosphine hydrochloride (TCEP). Because transition metals are needed in catalyzing peroxide formation from ascorbate, they also used membrane permeant and impermeant chelators for removal of the transition metals. The H_2O_2 scavengers completely protected the cells from ascorbate-induced death, indicating that H_2O_2 was the acting effector. The chelators were not protective, indicating the ascorbate-induced cell death was not due to chelatable trace redox-active metals. Superoxide dismutase was also tested and was not protective, because by dismutating superoxide radical, it produces H_2O_2 rather than degrades it [24]. A number of other radical or peroxide scavengers were tested later and by other investigators, as well as some antioxidants. All showed protective effects, confirming the central role of peroxide in the ascorbate-induced cell death [25–32].

Since the ascorbate-induced cell death was not due to chelatable redox-active metals, the investigators were curious as to what mediated the H_2O_2 formation from ascorbate. They found that the formation of H_2O_2 in the culture media was linearly correlated to the amount of ascorbate radical, the first product of ascorbate oxidation. The ascorbate radical formation, as well as concentrations of H_2O_2 formed, were dependent on the concentrations of ascorbate, the time of incubation, and importantly, a trace amount of serum proteins in the cell culture media [18]. At least 0.5% of serum was needed to mediate maximum production of H_2O_2 from ascorbate. The responsible proteins were not identified but were likely to be 10–30 kDa and to presumably contain metal centers [18]. In the presence of 0.5% serum, the formation of H_2O_2 did not require the presence of cells. When cells were present, H_2O_2 production was similar to either cancer cells or normal cells.

The cytotoxicity of ascorbate to cell lines was previously noted. However, because conditions used had a wide variety of concentrations and time courses that had no corresponding physiologic context, or used controls that contained no ascorbate at all, or used assays that ascorbate would interfere with, these studies were difficult to interpret in a clinically relevant way [33–37]. H_2O_2 generation by ascorbate oxidation in culture media was also known but was thought to be no more than an artifact of redox-active transition metals existing in the culture system [38,39], even though chelators had no effect [40,41]. Now, with new perspectives and data, Chen et al. realized that this "artifact" could actually be biologically and clinically relevant. The extracellular fluid in the body is basically a filtrate from the plasma, where as much as 20% of lower molecular weight serum proteins exist [42]. Although the responsible catalyst has not been identified, it is reasonable to postulate that it contains iron or copper or other redox-active metal center that catalyzes ascorbate oxidation. This metal center may be protected from chelating or even if chelated could still participate in the oxidation of ascorbate when ascorbate is at pharmacologic concentrations [43]. Or, even if the *in vitro* reaction to form H_2O_2 did not require cells, *in vivo* the cell surface proteins could harbor transition metals and provide catalyzation for ascorbate oxidation. Therefore, ascorbate radical and peroxide could form in tissues when pharmacologic concentrations of ascorbate were provided.

VALIDATION OF THE PRO-OXIDANT ACTION OF ASCORBATE *IN VIVO*

If ascorbate-induced cancer cell death is mediated by H_2O_2 formation, then it is of essential importance to know whether H_2O_2 is formed *in vivo* under clinically achievable conditions. As Chen et al. noted in their paper published in the *Proceedings of the National Academy of Sciences* 2005: "Based on the proposed reactions, if the predicted products are formed *in vivo*, then many next steps are justified, including determining molecular mechanisms of ascorbate action, isolation of proteins that mediate H_2O_2 formation, full characterization of ascorbate's preferential action on malignant but not normal cells, and animal and clinical trials. If the predicted reaction products are not formed *in vivo*, then a potential role of ascorbate in cancer treatment would require an entirely new explanation or may have to be discarded" [18].

They carried out the *in vivo* tests first in rats and then in tumor-bearing mice [19,44], with help from a chemist at the NIH, Kenneth Kirk, who synthesized a boronate fluorophore [45], namely PX1, for their *in situ* detection of H_2O_2 in the extracellular fluid obtained through microdialysis perfusion. First, they determined whether parenteral administration (intravenous or intraperitoneal) of ascorbate could give pharmacologic concentrations in the extracellular fluid, alongside those in the blood. Rats were given 0.25–0.5 g/kg body weight of ascorbate (similar to human doses) by tail vein injection (IV), injection into the peritoneal cavity (intraperitoneal [IP]), or by oral gavage. As expected, only the IV and IP doses were able to bring blood concentrations to millimolar range, about 60- to 100-fold higher than those from the oral doses. These millimolar concentrations were distributed to the extracellular fluids. Ascorbate radical was formed as an exponential function of ascorbate concentrations in the extracellular fluids from the tissue sites. H_2O_2 in the extracellular fluid was detected when ascorbate radical concentrations exceeded 100 nM, which only occurred with IV or IP administration [44]. In comparison, minimal to undetectable ascorbate radial was found in the blood even with the presence of >10 mM of ascorbate. H_2O_2 was always undetectable in the blood. Later, a higher dose of 4 g/kg IP injection was used in tumor-bearing mice, and similar results were found [19]. The IP dose increased ascorbate concentration to ~30 mM in the blood and in the extracellular fluids from both the tumor site and normal tissue site. These concentrations of ascorbate were greater than 150-fold higher compared to baseline and sustained for up to 3 hours. Ascorbate radical increased in the extracellular fluid from both tumor and normal tissue sites to greater than 500 nM, but blood radical concentrations were minimal (<50 nM). H_2O_2 was detected in both tumor and normal tissue extracellular fluids, achieving a plateau of 150 μM in 30 minutes after ascorbate IP injection and sustained for up to 3 hours, a level that could be cytotoxic to many cancer cells [19].

For the patient's safety, it is advantageous for the increase of the steady-state concentration of H_2O_2 in the extracellular milieu but not in the blood, after high-dose ascorbate administration, so that blood toxicity could be avoided. Several reasons could explain why ascorbate radical and H_2O_2 were not detected in the blood. First, the membrane of the red blood cell contains reducing activities that prevent ascorbate from losing an electron and forming ascorbate radical [46]. Therefore, a very low to undetectable level of ascorbate radical was found in the blood, lower than needed to form detectable steady-state concentrations of H_2O_2. Second, even if H_2O_2 was generated, it was instantly removed by catalase and glutathione peroxidase/reductase, which are enriched within red blood cells [47–51]. The rapid scavenging of H_2O_2 was also evident by spiking H_2O_2 solution into a sample of whole blood, and no H_2O_2 was detected [18].

Taken together, the data indicated that high-dose parenteral ascorbate is a peroxide delivery system for generation of sustainable ascorbate radical and H_2O_2 in the extracellular space, without accumulation of ascorbate radical or other reactive oxygen species in blood (Figure 2.2) [18,19,44]. This predicts effects on the tissue sites with low blood toxicity. The lack of blood toxicity is consistently observed in the clinical uses [10,52–63]. However, there is one contraindication for IV ascorbate use that has increased blood toxicity, and the contraindication is the deficiency in glucose-6-phosphate dehydrogenase (G6PD). In G6PD-deficient patients, high IV doses of ascorbate cause intravascular hemolysis [64,65]. This previously unexplained observation now can be explained by the formation of H_2O_2 from ascorbate. When the red blood cells remove H_2O_2, the intracellular glutathione (GSH) provides the needed reducing equivalents, and GSH is oxidized to glutathione disulfide (GSSG). To regenerate GSH from GSSG, the needed electron comes from NADPH, which is produced from the pentose phosphorylation pathway via G6PD catalyzation. If the G6PD enzyme activity is diminished, then the removal of H_2O_2 by the red blood cells will be impaired, and intravascular hemolysis results.

A number of independent animal studies were conducted using high-dose parenteral ascorbate to

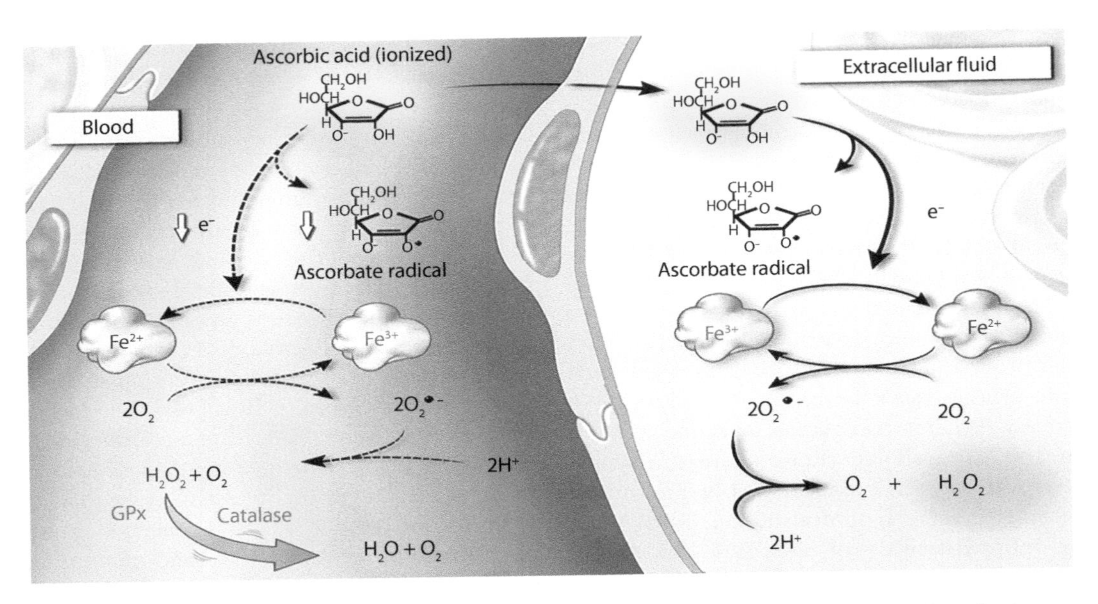

Figure 2.2. High-dose parenteral ascorbate is a prodrug to deliver ascorbate radical and hydrogen peroxide to the tissues. The parenteral (intravenous or intraperitoneal) administration achieves pharmacologic concentrations of ascorbate in the blood (left side). These high concentrations are distributed to extravascular spaces (right side). In the extracellular fluid, ascorbate is oxidized, donating an electron, and forming the ascorbate radical. The electron is passed to O_2 through a catalyst (unidentified, probably a protein) containing transition metal (Fe^{3+}/Fe^{2+} shown as an example), and subsequently forms superoxide radicals. H_2O_2 is generated through superoxide dismutation. These reactions are inhibited in the blood, first by membrane-reducing proteins on the red cells that prevent ascorbate radical formation. Even if H_2O_2 is formed, it is immediately degraded by the abundant catalase and peroxidase activities present in the blood. (From Chen, Q. et al. 2007 *Proc. Natl. Acad. Sci. USA* 104[21], 8749–8754. Reproduced with permission from the National Academy of Sciences, USA.)

treat a variety of cancers. The results consistently showed that high-dose parenteral ascorbate inhibited growth of xenografts of various types of aggressive cancers, including glioblastoma, pancreatic cancer, ovarian cancer, prostate cancer, hepatoma, colon cancer, sarcoma, mesothelioma, breast cancer, neuroblastoma, myeloid lymphoma, and many more [18,19,25–32,44,66]. Parenteral ascorbate has also been tested in combination with first-line therapies to various cancers, both *in vitro* and in rodent models. These studies from many research groups reported unequivocal data supporting additive to synergistic effects [27,55,67]. Now, the benefits of adding ascorbate to standard cancer therapies are under active clinical investigation. Clinical trials in multiple types of cancers have been conducted or are underway, as discussed in Chapter 7.

FURTHER DISCUSSION

The formation of radicals and peroxide in the tissues offers an attractive explanation as to why tight control of ascorbate concentrations occurs at physiologic conditions. When tight control is bypassed, radicals and other reactive oxygen species form and expose the tissues to oxidative stress. As the tight control mechanisms restore the physiologic ascorbate concentrations, H_2O_2 formation ceases. If the tight control mechanisms for ascorbate did not exist, there would be a risk of exposing tissues to constant high levels of H_2O_2. This may influence cell division and growth and have unwanted consequences [68–70]. In addition, ascorbate at physiologic concentrations is known to function as a reducing cofactor for a group of mental-centered enzymes, for example prolyl or lysyl hydroxylases, which are essential in collagen synthesis [71,72]. It is possible that high (pharmacologic) concentrations of ascorbate, if tight control did not exist, could react with a larger set of metalloproteins with higher Km, which may or may not be desirable under normal physiologic conditions. Therefore, the tight control mechanisms prevent tissue exposure to high concentrations of H_2O_2, as well as other possible reactions that normally do not happen by dietary ascorbate ingestion. In disease settings, such as in the context of the cancer milieu, temporary bypass of the tight control mechanisms allows episodic H_2O_2 formation, which selectively induces cancer cell death. The benefit of killing cancer cells outweighs the likelihood of harm to normal tissues. Indeed, little toxicities were found in cancer trials using high-dose intravenous ascorbate.

The pro-oxidant role of ascorbate is in contrast to its traditional antioxidant effect, but it is based on the same chemistry of ascorbate donating electrons and being oxidized. Under the condition of large-dose intravenous uses, the electron ascorbate lost is received by O_2 through transition metal ions and subsequent H_2O_2 generation is observed. This pro-oxidant effect provides new insights in using ascorbate in cancer treatment, as well as in other conditions that H_2O_2 may play a role, such as infections.

What makes cancer cells susceptible to ascorbate treatment but normal cells not is still an ongoing area of research with much interest. We postulate that ascorbate-induced peroxide has multiple targets in different parts of a cell, and different cells may respond to the challenge differently depending on the content. A few chapters in this book discuss several mechanisms of cellular responses to ascorbate treatment. Understanding the cellular mechanisms will help with understanding the reasons for sensitivity/resistance to ascorbate treatment and will guide combination therapy as well as patient selection.

REFERENCES

1. Cameron, E., Rotman, D. 1972 Ascorbic acid, cell proliferation, and cancer. *Lancet* 1(7749), 542. Epub 1972/03/04. PubMed PMID: 4110043
2. Cameron, E., Pauling, L. 1976 Supplemental ascorbate in the supportive treatment of cancer: Prolongation of survival times in terminal human cancer. *Proc. Natl. Acad. Sci. USA* 73(10), 3685–3689. Epub 1976/10/01. PubMed PMID: 1068480; PubMed Central PMCID: PMCPmc431183
3. Cameron, E., Campbell, A. 1974 The orthomolecular treatment of cancer. II. Clinical trial of high-dose ascorbic acid supplements in advanced human cancer. *Chem. Biol. Interact.* 9(4), 285–315. Epub 1974/10/01. doi: 0009-2797(74)90019-2 [pii]. PubMed PMID: 4430016
4. Cameron, E., Pauling, L. 1978 Supplemental ascorbate in the supportive treatment of cancer: Reevaluation of prolongation of survival times in terminal human cancer. *Proc. Natl. Acad. Sci. USA* 75(9), 4538–4542. Epub 1978/09/01. PubMed PMID: 279931; PubMed Central PMCID: PMCPmc336151
5. Creagan, E. T., Moertel, C. G., O'Fallon, J. R., Schutt, A. J., O'Connell, M. J., Rubin, J. et al. 1979 Failure of high-dose vitamin C (ascorbic acid) therapy to benefit patients with advanced cancer. A controlled trial. *N. Engl. J. Med.* 301(13), 687–690. Epub 1979/09/27. PubMed PMID: 384241
6. Moertel, C. G., Fleming, T. R., Creagan, E. T., Rubin, J., O'Connell, M. J., Ames, M. M. 1985 High-dose vitamin C versus placebo in the treatment of patients with advanced cancer who have had no prior chemotherapy. A randomized double-blind comparison. *N. Engl. J. Med.* 312(3), 137–141. Epub 1985/01/17. PubMed PMID: 3880867
7. Padayatty, S. J., Levine, M. 2000 Reevaluation of ascorbate in cancer treatment: Emerging evidence, open minds and serendipity. *J. Am. Coll. Nutr.* 19(4), 423–425. Epub 2000/08/30. PubMed PMID: 10963459
8. NBJ's Supplement Business Report 2008: An Anlysis of Markets, Trends, Competition and Strategy in the US Dietary Supplement Industry. 2008
9. Pauling, L. 1976 *Vitamin C, the Common Cold & the Flu*. Berkley Books.
10. Padayatty, S. J., Sun, A. Y., Chen, Q., Espey, M. G., Drisko, J., Levine, M. 2010 Vitamin C: Intravenous use by complementary and alternative medicine practitioners and adverse effects. *PLOS ONE* 5(7), e11414. Epub 2010/07/16. doi: 10.1371/journal.pone.0011414 [doi]. PubMed PMID: 20628650; PubMed Central PMCID: PMC2898816
11. Wittes, R. E. 1985 Vitamin C and cancer. *N. Engl. J. Med.* 312(3), 178–179. Epub 1985/01/17. PubMed PMID: 3965937
12. Levine, M. 1986 New concepts in the biology and biochemistry of ascorbic acid. *N. Engl. J. Med.* 314(14), 892–902. Epub 1986/04/03. PubMed PMID: 3513016
13. Levine, M., Dhariwal, K. R., Washko, P. W., Butler, J. D., Welch, R. W., Wang, Y. H. et al. 1991 Ascorbic acid and in situ kinetics: A new approach to vitamin requirements. *Am. J. Clin. Nutr.* 54(Suppl 6), 1157S–1162S. Epub 1991/12/01. PubMed PMID: 1962564
14. Levine, M., Conry-Cantilena, C., Wang, Y., Welch, R. W., Washko, P. W., Dhariwal, K. R. et al. 1996 Vitamin C pharmacokinetics in healthy volunteers: Evidence for a recommended

dietary allowance. *Proc. Natl. Acad. Sci. USA* 93(8), 3704–3709. Epub 1996/04/16. PubMed PMID: 8623000; PubMed Central PMCID: PMC39676

15. Levine, M., Wang, Y., Padayatty, S. J., Morrow, J. 2001 A new recommended dietary allowance of vitamin C for healthy young women. *Proc. Natl. Acad. Sci. USA* 98(17), 9842–9846. Epub 2001/08/16. doi: 10.1073/pnas.171318198 98/17/9842 [pii]. PubMed PMID: 11504949; PubMed Central PMCID: PMC55540
16. Graumlich, J. F., Ludden, T. M., Conry-Cantilena, C., Cantilena, L. R., Jr., Wang, Y., Levine, M. 1997 Pharmacokinetic model of ascorbic acid in healthy male volunteers during depletion and repletion. *Pharm. Res.* 14(9), 1133–1139. Epub 1997/11/05. PubMed PMID: 9327438
17. Padayatty, S. J., Sun, H., Wang, Y., Riordan, H. D., Hewitt, S. M., Katz, A. et al. 2004 Vitamin C pharmacokinetics: Implications for oral and intravenous use. *Ann. Intern. Med.* 140(7), 533–537. Epub 2004/04/08. doi: 140/7/533 [pii]. PubMed PMID: 15068981
18. Chen, Q., Espey, M. G., Krishna, M. C., Mitchell, J. B., Corpe, C. P., Buettner, G. R. et al. 2005 Pharmacologic ascorbic acid concentrations selectively kill cancer cells: Action as a pro-drug to deliver hydrogen peroxide to tissues. *Proc. Natl. Acad. Sci. USA* 102(38), 13604–9. Epub 2005/09/15. doi: 0506390102 [pii] 10.1073/pnas.0506390102. PubMed PMID: 16157892; PubMed Central PMCID: PMC1224653
19. Chen, Q., Espey, M. G., Sun, A. Y., Pooput, C., Kirk, K. L., Krishna, M. C. et al. 2008 Pharmacologic doses of ascorbate act as a prooxidant and decrease growth of aggressive tumor xenografts in mice. *Proc. Natl. Acad. Sci. USA* 105(32), 11105–9. doi: 10.1073/pnas.0804226105. PubMed PMID: 18678913; PubMed Central PMCID: PMC2516281
20. Corpe, C. P., Lee, J. H., Kwon, O., Eck, P., Narayanan, J., Kirk, K. L. et al. 2005 6-Bromo-6-deoxy-L-ascorbic acid: An ascorbate analog specific for Na+-dependent vitamin C transporter but not glucose transporter pathways. *J. Biol. Chem.* 280(7), 5211–5220. Epub 2004/12/14. doi: M412925200 [pii]10.1074/jbc.M412925200. PubMed PMID: 15590689
21. Wang, Y. H., Dhariwal, K. R., Levine, M. 1992 Ascorbic acid bioavailability in humans. Ascorbic acid in plasma, serum, and urine. *Ann. N. Y. Acad. Sci.* 669, 383–386. Epub 1992/09/30. PubMed PMID: 1444054
22. Levine, M., Dhariwal, K. R., Washko, P., Welch, R., Wang, Y. H., Cantilena, C. C. et al. 1992 Ascorbic acid and reaction kinetics in situ: A new approach to vitamin requirements. *J. Nutr. Sci. Vitaminol (Tokyo)*. Spec No, 169–172. Epub 1992/01/01. PubMed PMID: 1297733
23. Washko, P. W., Welch, R. W., Dhariwal, K. R., Wang, Y., Levine, M. 1992 Ascorbic acid and dehydroascorbic acid analyses in biological samples. *Anal. Biochem.* 204(1), 1–14. Epub 1992/07/01. doi: 0003-2697(92)90131-P [pii]. PubMed PMID: 1514674
24. Fridovich, I. 1995 Superoxide radical and superoxide dismutases. *Annu. Rev. Biochem.* 64, 97–112. doi: 10.1146/annurev.bi.64.070195.000525. PubMed PMID: 7574505
25. Du, J., Martin, S. M., Levine, M., Wagner, B. A., Buettner, G. R., Wang, S. H. et al. 2010 Mechanisms of ascorbate-induced cytotoxicity in pancreatic cancer. *Clin. Cancer Res.* 16(2), 509–520. Epub 2010/01/14. doi: 1078-0432.CCR-09-1713 [pii]10.1158/1078-0432.CCR-09-1713. PubMed PMID: 20068072; PubMed Central PMCID: PMC2807999
26. Ohno, S., Ohno, Y., Suzuki, N., Soma, G., Inoue, M. 2009 High-dose vitamin C (ascorbic acid) therapy in the treatment of patients with advanced cancer. *Anticancer Res.* 29(3), 809–815. Epub 2009/05/06. doi: 29/3/809 [pii]. PubMed PMID: 19414313
27. Verrax, J., Calderon, P. B. 2009 Pharmacologic concentrations of ascorbate are achieved by parenteral administration and exhibit antitumoral effects. *Free. Radic. Biol. Med.* 47(1), 32–40. Epub 2009/03/04. doi: 10.1016/j.freeradbiomed.2009.02.016. PubMed PMID: 19254759
28. Pollard, H. B., Levine, M. A., Eidelman, O., Pollard, M. 2010 Pharmacological ascorbic acid suppresses syngeneic tumor growth and metastases in hormone-refractory prostate cancer. *In Vivo*. 24(3), 249–255. Epub 2010/06/18. doi: 24/3/249 [pii]. PubMed PMID: 20554995
29. Takemura, Y., Satoh, M., Satoh, K., Hamada, H., Sekido, Y., Kubota, S. 2010 High dose of ascorbic acid induces cell death in mesothelioma cells. *Biochem. Biophys. Res. Commun.* 394(2), 249–253. Epub 2010/02/23. doi: S0006-291X(10)00212-3 [pii]10.1016/j.bbrc.2010.02.012. PubMed PMID: 20171954
30. Ullah, M. F., Khan, H. Y., Zubair, H., Shamim, U., Hadi, S. M. 2010 The antioxidant ascorbic acid

mobilizes nuclear copper leading to a prooxidant breakage of cellular DNA: Implications for chemotherapeutic action against cancer. *Cancer Chemother. Pharmacol.* 67(1), 103–110. Epub 2010/03/10. doi: 10.1007/s00280-010-1290-4. PubMed PMID: 20213077

31. Gilloteaux, J., Jamison, J. M., Neal, D. R., Loukas, M., Doberzstyn, T., Summers, J. L. 2010 Cell damage and death by autoschizis in human bladder (RT4) carcinoma cells resulting from treatment with ascorbate and menadione. *Ultrastruct. Pathol.* 34(3), 140–160. Epub 2010/05/12. doi: 10.3109/01913121003662304. PubMed PMID: 20455663
32. Deubzer, B., Mayer, F., Kuci, Z., Niewisch, M., Merkel, G., Handgretinger, R. et al. 2010 H(2) O(2)-mediated cytotoxicity of pharmacologic ascorbate concentrations to neuroblastoma cells: Potential role of lactate and ferritin. *Cell Physiol. Biochem.* 25(6), 767–774. Epub 2010/06/01. doi: 10.1159/000315098. PubMed PMID: 20511723
33. Sakagami, T., Satoh, K., Ishihara, M., Sakagami, H., Takeda, F., Kochi, M. et al. 2000 Effect of cobalt ion on radical intensity and cytotoxic activity of antioxidants. *Anticancer Res.* 20(5A), 3143–3150. Epub 2000/11/04. PubMed PMID: 11062735
34. Han, S. S., Kim, K., Hahm, E. R., Lee, S. J., Surh, Y. J., Park, H. K. et al. 2004 L-ascorbic acid represses constitutive activation of NF-kappaB and COX-2 expression in human acute myeloid leukemia, HL-60. *J. Cell Biochem.* 93(2), 257–270. Epub 2004/09/16. doi: 10.1002/jcb.20116. PubMed PMID: 15368354
35. Park, S., Han, S. S., Park, C. H., Hahm, E. R., Lee, S. J., Park, H. K. et al. 2004 L-Ascorbic acid induces apoptosis in acute myeloid leukemia cells via hydrogen peroxide-mediated mechanisms. *Int. J. Biochem. Cell. Biol.* 36(11), 2180–2195. Epub 2004/08/18. doi: 10.1016/j.biocel.2004.04.005 S1357272504001591 [pii]. PubMed PMID: 15313465
36. Koh, W. S., Lee, S. J., Lee, H., Park, C., Park, M. H., Kim, W. S. et al. 1998 Differential effects and transport kinetics of ascorbate derivatives in leukemic cell lines. *Anticancer Res.* 18(4A), 2487–2493. Epub 1998/08/15. PubMed PMID: 9703897
37. Calderon, P. B., Cadrobbi, J., Marques, C., Hong-Ngoc, N., Jamison, J. M., Gilloteaux, J. et al. 2002 Potential therapeutic application of the association of vitamins C and K3 in cancer treatment. *Curr. Med. Chem.* 9(24), 2271–2285. Epub 2002/12/10. PubMed PMID: 12470246
38. Arakawa, N., Nemoto, S., Suzuki, E., Otsuka, M. 1994 Role of hydrogen peroxide in the inhibitory effect of ascorbate on cell growth. *J. Nutr. Sci. Vitaminol (Tokyo).* 40(3), 219–227. PubMed PMID: 7965211
39. Sestili, P., Brandi, G., Brambilla, L., Cattabeni, F., Cantoni O. 1996 Hydrogen peroxide mediates the killing of U937 tumor cells elicited by pharmacologically attainable concentrations of ascorbic acid: Cell death prevention by extracellular catalase or catalase from cocultured erythrocytes or fibroblasts. *J. Pharmacol. Exp. Ther.* 277(3), 1719–1725. PubMed PMID: 8667243
40. Buettner, G. R. 1986 Ascorbate autoxidation in the presence of iron and copper chelates. *Free. Radic. Res. Commun.* 1(6), 349–353. PubMed PMID: 2851502
41. Clement, M. V., Ramalingam, J., Long, L. H., Halliwell, B. 2001 The in vitro cytotoxicity of ascorbate depends on the culture medium used to perform the assay and involves hydrogen peroxide. *Antioxid. Redox. Signal.* 3(1), 157–163. doi: 10.1089/152308601750100687. PubMed PMID: 11291594
42. Weinberger, A., Simkin, P. A. 1989 Plasma proteins in synovial fluids of normal human joints. *Semin. Arthritis. Rheum.* 19(1), 66–76. PubMed PMID: 2672342
43. Halliwell, B. 1990 How to characterize a biological antioxidant. *Free. Radic. Res. Commun.* 9(1), 1–32. PubMed PMID: 2159941
44. Chen, Q., Espey, M. G., Sun, A. Y., Lee, J. H., Krishna, M. C., Shacter, E. et al. 2007 Ascorbate in pharmacologic concentrations selectively generates ascorbate radical and hydrogen peroxide in extracellular fluid in vivo. *Proc. Natl. Acad. Sci. USA* 104(21), 8749–8754. Epub 2007/05/16. doi: 0702854104 [pii]10.1073/pnas.0702854104. PubMed PMID: 17502596; PubMed Central PMCID: PMC1885574
45. Miller, E. W., Albers, A. E., Pralle, A., Isacoff, E. Y., Chang, C. J. 2005 Boronate-based fluorescent probes for imaging cellular hydrogen peroxide. *J. Am. Chem. Soc.* 127(47), 16652–9. doi: 10.1021/ja054474f. PubMed PMID: 16305254; PubMed Central PMCID: PMCPMC1447675
46. Wang, X., Liu, J., Yokoi, I., Kohno, M., Mori, A. 1992 Direct detection of circulating free radicals in the rat using electron spin resonance spectrometry. *Free Radic. Biol. Med.* 12(2), 121–126. PubMed PMID: 1313773

47. Kakinuma, K., Yamaguchi, T., Kaneda, M., Shimada, K., Tomita, Y., Chance, B. 1979 A determination of H_2O_2 release by the treatment of human blood polymorphonuclear leukocytes with myristate. *J. Biochem.* 86(1), 87–95. PubMed PMID: 479132
48. Chance, B., Sies, H., Boveris, A. 1979 Hydroperoxide metabolism in mammalian organs. *Physiol. Rev.* 59(3), 527–605. doi: 10.1152/physrev.1979.59.3.527. PubMed PMID: 37532
49. Brown, J. M., Grosso, M. A., Terada, L. S., Beehler, C. J., Toth, K. M., Whitman, G. J. et al. 1989 Erythrocytes decrease myocardial hydrogen peroxide levels and reperfusion injury. *Am. J. Physiol.* 256(2 Pt 2), H584–H588. doi: 10.1152/ajpheart.1989.256.2.H584. PubMed PMID: 2916691
50. Guemouri, L., Artur, Y., Herbeth, B., Jeandel, C., Cuny, G., Siest, G. 1991 Biological variability of superoxide dismutase, glutathione peroxidase, and catalase in blood. *Clin. Chem.* 37(11), 1932–1937. PubMed PMID: 1934468
51. Motoyama, S., Saito, S., Inaba, H., Kitamura, M., Minamiya, Y., Suzuki, H. et al. 2000 Red blood cells attenuate sinusoidal endothelial cell injury by scavenging xanthine oxidase-dependent hydrogen peroxide in hyperoxic perfused rat liver. *Liver* 20(3), 200–208. PubMed PMID: 10902969
52. Schoenfeld, J. D., Sibenaller, Z. A., Mapuskar, K. A., Wagner, B. A., Cramer-Morales, K. L., Furqan, M. et al. 2017 O2- and H_2O_2-Mediated disruption of Fe Metabolism causes the differential susceptibility of NSCLC and GBM cancer cells to pharmacological ascorbate. *Cancer Cell.* 31(4), 487–500. doi: 10.1016/j.ccell.2017.02.018. PubMed PMID: 28366679
53. Polireddy, K., Dong, R., Reed, G., Yu, J., Chen, P., Williamson, S. et al. 2017 High dose parenteral ascorbate inhibited pancreatic cancer growth and metastasis: Mechanisms and a phase I/IIa study. *Sci. Rep.* 7(1), 17188. doi: 10.1038/s41598-017-17568-8. PubMed PMID: 29215048; PubMed Central PMCID: PMCPMC5719364
54. Du, J., Cieslak, J. A. 3rd, Welsh, J. L., Sibenaller, Z. A., Allen, B. G., Wagner, B. A. et al. 2015 Pharmacological ascorbate radiosensitizes pancreatic cancer. *Cancer Res.* 75(16), 3314–3326. Epub 2015/06/18. doi: 10.1158/0008-5472.can-14-1707. PubMed PMID: 26081808; PubMed Central PMCID: PMCPmc4537815
55. Ma, Y., Chapman, J., Levine, M., Polireddy, K., Drisko, J., Chen, Q. 2014 High-dose parenteral ascorbate enhanced chemosensitivity of ovarian cancer and reduced toxicity of chemotherapy. *Sci. Transl. Med.* 6(222), 222ra18. doi: 10.1126/scitranslmed.3007154. PubMed PMID: 24500406
56. Nauman, G., Gray, J. C., Parkinson, R., Levine, M., Paller, C. J. 2018 Systematic review of intravenous ascorbate in cancer clinical trials. *Antioxidants (Basel)* 7(7). doi: 10.3390/antiox7070089. PubMed PMID: 30002308; PubMed Central PMCID: PMCPMC6071214
57. Alexander, M. S., Wilkes, J. G., Schroeder, S. R., Buettner, G. R., Wagner, B. A., Du, J. et al. 2018 Pharmacologic ascorbate reduces radiation-induced normal tissue toxicity and enhances tumor radiosensitization in pancreatic cancer. *Cancer Res.* 78(24), 6838–6851. doi: 10.1158/0008-5472.CAN-18-1680. PubMed PMID: 30254147; PubMed Central PMCID: PMCPMC6295907
58. Takahashi, H., Mizuno, H., Yanagisawa, A. 2012 High-dose intravenous vitamin C improves quality of life in cancer patients. *Personalized Med. Universe* 1(1), 49–53.
59. Ou, J., Zhu, X., Lu, Y., Zhao, C., Zhang, H., Wang, X. et al. 2017 The safety and pharmacokinetics of high dose intravenous ascorbic acid synergy with modulated electrohyperthermia in Chinese patients with stage III-IV non-small cell lung cancer. *Euro. J. Pharm. Sci.* 109, 412–418. Epub 2017/08/30. doi: 10.1016/j.ejps.2017.08.011. PubMed PMID: 28847527
60. Hoffer, L. J., Robitaille, L., Zakarian, R., Melnychuk, D., Kavan, P., Agulnik, J. et al. 2015 High-dose intravenous vitamin C combined with cytotoxic chemotherapy in patients with advanced cancer: A phase I-II clinical trial. *PLOS ONE* 10(4), e0120228. Epub 2015/04/08. doi: 10.1371/journal.pone.0120228. PubMed PMID: 25848948; PubMed Central PMCID: PMCPmc4388666
61. Kawada, H., Sawanobori, M., Tsuma-Kaneko, M., Wasada, I., Miyamoto, M., Murayama, H. et al. 2014 Phase I clinical trial of intravenous L-ascorbic acid following salvage chemotherapy for relapsed B-cell non-Hodgkin's lymphoma. *Tokai J. Exp. Clin. Med.* 39(3), 111–115. Epub 2014/09/25. PubMed PMID: 25248425
62. Stephenson, C. M., Levin, R. D., Spector, T., Lis, C. G. 2013 Phase I clinical trial to evaluate the safety, tolerability, and pharmacokinetics of high-dose intravenous ascorbic acid in patients with advanced cancer. *Cancer Chemother. Pharmacol.* 72(1), 139–146. Epub 2013/05/15. doi: 10.1007/s00280-013-2179-9. PubMed PMID: 23670640; PubMed Central PMCID: PMCPmc3691494

63. Held, L. A., Rizzieri, D., Long, G. D., Gockerman, J. P., Diehl, L. F., de Castro, C. M. et al. 2013 A Phase I study of arsenic trioxide (Trisenox), ascorbic acid, and bortezomib (Velcade) combination therapy in patients with relapsed/refractory multiple myeloma. *Cancer Invest.* 31(3), 172–176. Epub 2013/02/15. doi: 10.3109/07357907.2012.756109. PubMed PMID: 23406188
64. Cameron, E., Campbell, A., Jack, T. 1975 The orthomolecular treatment of cancer. III. Reticulum cell sarcoma: Double complete regression induced by high-dose ascorbic acid therapy. *Chem. Biol. Interact.* 11(5), 387–393. Epub 1975/11/01. doi: 0009-2797(75)90007-1 [pii]. PubMed PMID: 1104207
65. Rees, D. C., Kelsey, H., Richards, J. D. 1993 Acute haemolysis induced by high dose ascorbic acid in glucose-6-phosphate dehydrogenase deficiency. *BMJ* 306(6881), 841–842. doi: 10.1136/bmj.306.6881.841. PubMed PMID: 8490379; PubMed Central PMCID: PMCPMC1677333
66. Wei, Y., Song, J., Chen, Q., Xing, D. 2012 Enhancement of photodynamic antitumor effect with pro-oxidant ascorbate. *Lasers Surg. Med.* 44(1), 69–75. Epub 2012/01/17. doi: 10.1002/lsm.21157 [doi]. PubMed PMID: 22246986
67. Espey, M. G., Chen, P., Chalmers, B., Drisko, J., Sun, A. Y., Levine, M. et al. 2011 Pharmacologic ascorbate synergizes with gemcitabine in preclinical models of pancreatic cancer. *Free. Radic. Biol. Med.* 50(11), 1610–1619. Epub 2011/03/16. doi: 10.1016/j.freeradbiomed.2011.03.007. PubMed PMID: 21402145; PubMed Central PMCID: PMCPmc3482496
68. Stone, J. R., Yang, S. 2006 Hydrogen peroxide: A signaling messenger. *Antioxid. Redox. Signal.* 8(3-4), 243–270. Epub 2006/05/09. doi: 10.1089/ars.2006.8.243 [doi]. PubMed PMID: 16677071
69. Miyoshi, N., Oubrahim, H., Chock, P. B., Stadtman, E. R. 2006 Age-dependent cell death and the role of ATP in hydrogen peroxide-induced apoptosis and necrosis. *Proc. Natl. Acad. Sci. USA* 103(6), 1727–1731. Epub 2006/01/31. doi: 0510346103 [pii] 10.1073/pnas.0510346103. PubMed PMID: 16443681; PubMed Central PMCID: PMC1413652
70. Rhee, S. G. 2006 Cell signaling. H_2O_2, a necessary evil for cell signaling. *Science* 312(5782), 1882–1883. doi: 10.1126/science.1130481. PubMed PMID: 16809515
71. Englard, S., Seifter, S. 1986 The biochemical functions of ascorbic acid. *Annu. Rev. Nutr.* 6, 365–406. doi: 10.1146/annurev.nu.06.070186.002053. PubMed PMID: 3015170
72. Bruegge, K., Jelkmann, W., Metzen, E. 2007 Hydroxylation of hypoxia-inducible transcription factors and chemical compounds targeting the HIF-alpha hydroxylases. *Curr. Med. Chem.* 14(17), 1853–1862. Epub 2007/07/14. PubMed PMID: 17627521

CHAPTER THREE

Pharmacologic Ascorbate Influences Multiple Cellular Pathways Preferentially in Cancer Cells

Qi Chen, Kishore Polireddy, Ping Chen, Ramesh Balusu, Tao Wang, and Ruochen Dong

CONTENTS

INTRODUCTION

As high-dose intravenous ascorbate is noted to preferentially kill cancer cells but spare normal cells, it is not very clear how it does the selective killing. The critical underlying mechanism of action is that pharmacologic concentrations of ascorbate generate hydrogen peroxide (H_2O_2) [1–3]. Catalase, an enzyme that neutralizes H_2O_2, rescued ascorbate-mediated cell death in cancer cells. The evidence from *in vitro* studies was confirmed in animal model systems. Parenteral (intraperitoneal or intravenous) administration of ascorbate results in millimolar ascorbate concentrations in the bloodstream. These high ascorbate concentrations are distributed to the extracellular spaces. In the extracellular spaces, ascorbate loses an electron and generates ascorbate radical [2]. The ascorbate radical loses another electron and ascorbate becomes oxidized. The electrons are passed to O_2 molecules that then results in generation of superoxide radical (O_2^-). Dismutation of superoxide radical leads to H_2O_2 formation. H_2O_2 acts as the effective molecule and induces cancer cell death, inhibiting tumor growth and metastasis [1–3]. Pharmacologic ascorbate regulates the tumor inhibitory effects by inducing diverse biological responses that differently influence cancer cells and normal cells. Recent studies using multiple tumor models suggested that the selectivity is related to tumor cells' abnormal energy metabolism and iron metabolism [2,4–6]. This chapter discusses several mechanisms that lead to the preferential cancer cell inhibition. These mechanisms include but may not be limited to cancer cell iron metabolism, DNA damage, depletion of cancer cellular energy sources, and inhibition of microtubule dynamics. The multiple mechanisms are all related to peroxide-induced oxidative stress generated by pharmacologic ascorbate, and work in concert to ensure selective inhibition in cancer cells.

PEROXIDE INSIDE OR OUTSIDE THE CELL

As discussed in Chapter 2, it is the dramatically increased extracellular ascorbate concentration that forms H_2O_2 [2], and H_2O_2 can permeate the cell membrane and influence molecules inside the cell [1]. Downstream reactive oxygen species (ROS) can also form through Fenton chemistry. Exogenously added catalase or glutathione is preventive of ascorbate-induced cell death [1,7]. Also, when catalase is overexpressed in the cytoplasm through adenoviral gene transfection, it also rescues ascorbate-mediated cell death, in various cancer cell lines [8,9]. This indicates that the critical target of H_2O_2 is likely to be inside the cell, even if H_2O_2 is formed outside the cell.

There lies another possibility, as revealed by recent research, that cancer cells with abnormally increased uptake of ascorbate might be more sensitive, because excessive ascorbate can form ROS inside the cancer cell where pliable transition metal (such as iron) is available. Hong et al. showed that sodium-dependent vitamin C transporter 2 (SVCT2) sensitized breast cancer cells by increasing intracellular ascorbate and ROS; therefore, they proposed that SVCT2 can serve as a biomarker for sensitivity to ascorbate treatment in breast cancer patients [10]. Lv et al. investigated liver cancer (hepatocellular carcinoma) and found that the subpopulation of cancer cells with stem-like properties (or termed as cancer stem cells, CSCs) had upregulated SVCT2 and was more sensitive to ascorbate treatment [11]. Schoenfeld et al. postulated that, because of the known elevated endogenous ROS levels produced by cancer mitochondria, the cancer cells have increased uptake of iron through elevated transferrin receptor (TfR) expression [9]. In cancer cells with upregulated SVCT2, ascorbate concentration is increased. The increased labile iron and ascorbate inside the cell favor the production of peroxide, which in turn enhances iron uptake and more peroxide formation from intracellular ascorbate [9]. This positive feedback loop of iron uptake and ascorbate-induced ROS formation results in catastrophic consequences in cancer cells. Because normal cells do not have endogenously elevated iron uptake, this ROS formation feedback loop does not begin.

These observations postulate ROS formation inside certain cancer cells when ascorbate and iron uptake is enhanced, adding an aspect to the previous studies suggesting ROS formation outside the cell. Taken together, pharmacologic ascorbate forms H_2O_2 both outside the cells and inside the cells under some conditions. The elevated formation of ROS inside the cells partially explains why some cancer cells are more sensitive. However, a large part of the sensitivity still depends on how the cell responds to the H_2O_2 challenge. After all, cytotoxic levels of H_2O_2 are formed by pharmacologic ascorbate at tissue sites, outside the cells, regardless of tumor or normal [2], and whether H_2O_2 is formed outside or inside the cells, it can exert effects on molecules inside the cell.

INDUCTION OF DNA DAMAGE AND INHIBITION OF DNA REPAIR MACHINERY IN CANCER CELLS

The pro-oxidant molecule H_2O_2 is an important DNA reactive species and induces oxidative DNA damage. Fast dividing cells (such as cancer cells) are more susceptible to DNA damaging agent, and that makes it reasonable that many chemotherapeutics drugs are DNA damaging agents (e.g., platinum-based drugs cisplatin and carboplatin). Oxidative DNA damage includes single-stranded breaks (SSBs), double-stranded breaks (DSBs), base damage such as at apurinic or apyrimidinic sites, and inter- and intrastrand cross-links [12]. DNA damage is repaired by endogenous DNA repair machineries including homologous and nonhomologous repair. Unrepaired SSBs can result in DSBs, and severe DNA damage failed repair induces cell death. Ma et al. showed that ascorbate at millimolar concentrations caused DSBs in ovarian cancer cells resulting in cell death [4]. When ascorbate was combined with the DNA alkylating agent carboplatin, DNA damage increased, and cell death increased. Catalase, which neutralizes H_2O_2, rescued ascorbate-mediated DNA damage and reduced cell death [4]. Similar results were reproducible in neuroblastoma cells, that DNA DSB was evident in ascorbate-treated neuroblastoma cells, and when the DNA repair enzyme PARP was inhibited, cell death increased [4,13]. In another study, ascorbate in millimolar concentrations (5 mM) in combination with radiation killed more glioblastoma cells than radiation alone, by increasing DNA DSB [14].

A complex network of signaling pathways is altered when cells are exposed to DNA damaging agents [15]. Like other signaling pathways, a DNA damage response (DDR) signaling pathway consists of sensors, transducers, and effectors [16]. DNA damage sensors are the proteins that directly recognize aberrant DNA structures. Mre11-Rad50-Nbs1 (MRN) complex is the key sensor of DNA damage in mammalian cells; it activates ataxia-telangiectasia mutated (ATM) and ATM- and Rad3-related (ATR) kinases, two key transducers of the complex DDR network signaling. Pharmacologic ascorbate activates ATM in a concentration- and time-dependent manner by inducing ATM phosphorylation, and this phosphorylation can be rescued by catalase [4,13]. Following the initial activation, ATM triggers phosphorylation of histone 2Ax (H2Ax), which is a critical event for accumulation of numerous DNA repair proteins and chromatin-remodeling complexes around the DSBs [4,14]. Chk2, another downstream effector of ATM and ART, was also activated by ascorbate treatment [13]. It is also proposed that other downstream targets of ATM and ATR kinases (e.g., BRCA1/2, and p53) [17] are influenced by ascorbate treatment, which are primarily involved in a broad spectrum of cellular processes important for genomic stability and influence cell survival, cell cycle, apoptosis, and senescence [18,19].

After the DDR signaling pathways activate the DNA repair machinery in the cell, DSBs are repaired by two distinct pathways such as homologous recombination (HR) and nonhomologous end joining (NHEJ). HR is the most used mechanism in which genetic material is exchanged between sister chromatids to repair the damaged DNA without loss of nucleotides. During HR, the enzymes Rad51 and Dmc1 catalyze pairing and shuffling of homologous DNA sequences in mammalian cells, leading to precise repair of the damaged sites. This process is enhanced by breast tumor suppressor BRCA1/2 [20]. During NHEJ, broken ends are brought together and rejoined by DNA ligation, generally with the loss of one or more nucleotides at the site of joining; hence, it is an error-prone DNA repair mechanism. The protein Ku heterodimer (Ku70 and Ku80) recognizes DSBs and acts as a scaffold to recruit the other NHEJ factors, such as DNA-PKcs, x-ray cross complementing protein 4, DNA ligase IV, XRCC4-like factor, and aprataxin-and-PNK-like factor, to DSBs to complete the ligation process [21]. Recent data showed that pharmacologic ascorbate suppresses the expression of HR repair proteins including BRCA1, BRCA2, and RAD51, thus leading to HR deficiency and sensitizing the *BRCA1/2* wild-type epithelial ovarian cancer cells to PARP inhibition [22]. Meanwhile, in the presence of HR deficiency, pharmacologic ascorbate also impeded the NHEJ pathway, leading to DNA repair deficiency [22].

Thus, pharmacologic ascorbate not only damages cancer cell DNA, but it also suppresses DNA repairing machinery. This makes it mechanistically plausible to combine ascorbate not only with DNA damaging chemotherapeutics (such as platinum-based drugs), but also with PARP inhibitor—one of the most exciting new classes of oncology drugs that has transformed the management of ovarian cancer. An early phase clinical trial in ovarian cancer patients combing high-dose intravenous ascorbate (IVC) and the standard carboplatin/paclitaxel proved the safety and feasibility, achieved improved toxicity profiles in the patients, and suggested prolonged disease progression-free survival (PFS) [4]. PARP inhibitors (such as olaparib, rucaparib, and niraparib) exhibit good efficacy and clinical benefits in women carrying *BRCA1/2*-mutated ovarian cancers. However, the majority of women who have ovarian or breast cancer (~80%) do not have germline or somatic *BRCA1/2* mutation, to whom PARP inhibitors alone have very limited efficacy. In a mouse model, adding pharmacologic ascorbate treatment significantly sensitized a *BRCA1/2* wild-type ovarian cancer to olaparib treatment [22]. The combination treatment of ascorbate and PARP inhibitor provides a novel and promising therapeutic option for patients with cancers not responding to PARP inhibition alone, and it is worth clinical investigation. Such a strategy could be applied to a variety of heterogeneous and hard-to-treat malignancies, such as breast, pancreatic, and prostate cancers, where BRCA1, BRCA2 or other HR repair proteins are instrumental in the repair of DNA DSBs, and the potential of both PARP inhibitors and ascorbate has not been fully exploited.

DEPLETION OF ATP IN CANCER CELLS

H_2O_2 can cause rapid depletion of cellular energy stores and deplete NAD^+ and ATP in cells [23,24]. Chen et al. therefore first hypothesized that pharmacologic ascorbate can deplete ATP in sensitive cancer cells and result in cell death [2]. There were three mechanisms that they proposed through which cellular ATP can be depleted [2]: (1) The ascorbate-induced H_2O_2 damages DNA, and therefore, PARP is activated, intending to repair. Activated PARP uses NAD^+ as a substrate in synthesizing poly-ADP-ribose, and thereby, the availability of NAD^+ decreases for NADH formation, and this consequently limits ATP synthesis. (2) Glutathione (GSH) is oxidized to glutathione disulfate (GSSG) as a cellular defense against increased H_2O_2. To reduce

GSSG back to GSH, NADPH is needed as the reducing force. NADPH is provided by the pentose phosphorylation from glucose. Therefore, the flow of glucose usage shifts to reduce $NADP^+$ to NADPH, and glycolysis is inhibited and ATP production is decreased. (3) H_2O_2 may directly damage mitochondria, especially the ATP synthase, so that ATP production falls. Known as the Warburg effect [2,4,5,25], there are fundamental differences in energy metabolism between cancer and normal cells. Relative to normal cells, some cancer cells tend to use glycolysis for ATP production, rather than the more efficient oxidative phosphorylation pathway. It is believed that cancer cells prefer glycolysis because it is a rapid way to produce ATP and it promotes flux into biosynthetic pathways, which are needed to support the cancer cell's fast growth and division [26]. Nevertheless, compared with oxidative phosphorylation, ATP generation by glycolysis is inefficient (2 from glycolysis versus 36 ATPs from the TCA cycle from one glucose molecule). Therefore, cancer cells that are glycolysis dependent may be particularly sensitive to pharmacologic ascorbate concentrations, compared with cells that use oxidative phosphorylation [4]. These hypotheses are now validated by many studies.

First, studies have found that pharmacologic concentrations of ascorbate deplete ATP levels in various sensitive cancer cell lines but not in normal cells [4,5,8,13,27–29]. In neuroblastoma cells, Ma et al. [13] showed in detail that ascorbate-generated H_2O_2 damaged DNA, and accordingly, PARP was activated as the cell's intention to repair the damaged DNA. Because PARP uses NAD^+ as a substrate to form poly-ADP-ribose (PAR), a depletion in NAD^+ resulted in ascorbate-treated cancer cells. Consequently, the activity of glyceraldehyde 3-phosphate dehydrogenase (GAPDH), a critical enzyme in the glycolysis pathway, was inhibited, and reduced glycolysis flux resulted, leading to cell death. NAD^+ supplementation prevented ATP depletion and prevented cell death. ATP supplementation also prevented cell death. These studies provided unequivocal evidence supporting the hypothesis.

PARP inhibitors have been shown to rescue cellular ATP depletion [30]. Since ascorbate depletes NAD^+ and ATP in cancer cells through PARP activation, one would suspect that the PARP inhibitor will rescue the ATP drop caused by ascorbate and thus counteract with ascorbate in inducing cancer cell death. This raises a clinical concern for using ascorbate and PARP inhibitors together in oncologic patients, especially now that PARP inhibitors are now first-line anticancer drugs for women with heavily pretreated or recurrent ovarian cancer who harbor *BRCA* gene mutations. PARP inhibitors are also used in treating breast cancer, prostate cancer, and other *BRCA*-associated cancers. For this reason, it is important to answer the question whether ascorbate counteracts with PARP inhibitors.

Therefore, studies have been carried out combining PARP inhibitors with high-dose parenteral ascorbate in preclinical tumor models. As expected, PARP inhibitors protected NAD^+ and ATP levels [13]. However, PARP inhibitor did not prevent ascorbate-induced cell death despite the preserved NAD^+ and ATP levels [13,22]. This is presumably because PARP inhibition defects DNA repair and thus increases fatal DNA double-strand damage, regardless of ATP levels. Indeed, other studies have shown that PARP inhibition does not prevent oxidant-induced cell death but instead changes the mechanism of cell death [31].

These related but different mechanisms each play important roles in ascorbate-induced cancer cell death depending on the content. When a cancer cell has intact DNA repair machinery, it repairs DNA damage at the cost of NAD^+, resulting in a reduced glycolysis flux and ATP production, leading to cell death. When the PARP-dependent DNA repair machinery is inhibited, although ATP and NAD^+ are preserved, excessive DNA damage could accumulate and lead to cell death (Figure 3.1). Ideally, complementary but distinct cytotoxic mechanisms could reduce the emergence of treatment resistance. Also, in cancers that use alternative energy metabolism pathways other than glucose, combining a PARP inhibitor with ascorbate treatment could enhance effectiveness. As a further advantage, the dose of the PARP inhibitor could be reduced to minimize associated toxicities.

The ascorbate-induced loss of GAPDH activity has two possible explanations. First, GAPDH activity is directly influenced by NAD^+ levels. Second, ascorbate may oxidize the GAPDH active site (cysteine 152, C152) and lower GAPDH activity. Interestingly, a previous study showed that the C152 of GAPDH was only reversibly affected by pharmacologic ascorbate and underwent S-glutathionylation. The irreversibly

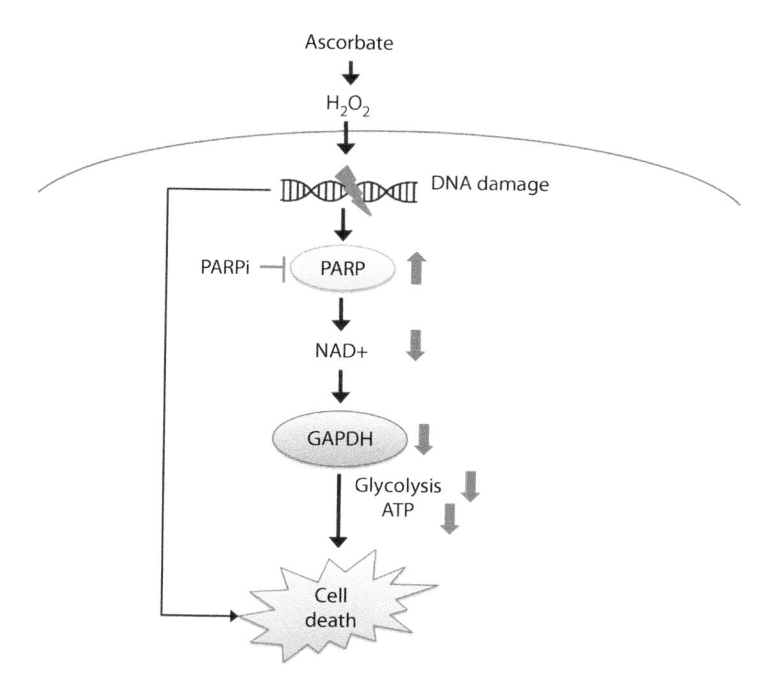

Figure 3.1. Pharmacologic ascorbate damages cancer cell DNA and inhibits ATP production. Through generation of H_2O_2, pharmacologic ascorbate damages cell DNA. PARP is subsequently activated as the cell's attempt to repair DNA damage. Activated PARP depletes NAD+, which is a co-enzyme for GAPDH activity. As GAPDH activity is inhibited, glycolysis flow is reduced and ATP is depleted in cancer cells, leading to cell death. In the absence of PARP activation, for example, using a PARP inhibitor (PARPi), fatal DNA damage accumulates and leads to cell death. (GAPDH, glyceraldehyde 3-phosphate dehydrogenase; PARP, poly-ADP-ribose polymerase.)

oxidized form, Cys-SO_3H, which meant to reduce GAPDH activity, was not detected [5]. Glutathionylation effects on protein activity are not completely known, but one of its roles is that it protects the cysteine amino acids from becoming further oxidized. These data indicate the oxidation at the active site of GAPDH may not be the mechanism. Meanwhile, as supplementation of NAD^+ prevents ascorbate-induced ATP loss in cancer cells, it is more likely that the loss of GAPDH activity in ascorbate-treated cells is due to the loss of NAD^+.

Although studies have not ruled out a role for mitochondria in ascorbate-induced cancer cell death, the mitochondrial contributions are likely to be limited. Studies have shown that mitochondrial DNA-depleted cancer cells (ρ0 cells) that lack functional respiratory chains have the same sensitivity to ascorbate as their mitochondria-intact parent cells [32]. Also, transfection and overexpression of catalase in mitochondria did not rescue the cell death under ascorbate treatment, whereas cytoplasmic overexpression of catalase did rescue the cells [32].

KRAS mutations account for approximately 15% of all human tumors. Recent efforts have been made to identify drugs that can potentially target this nearly undruggable target [33–35]. A recent study [5] suggested that oncogenic *KRAS/BRAF* sensitize colon cancer cells to ascorbate treatment, and suggested the underlying reason is because *KRAS/BRAF* mutations upregulate expression of GLUT1, the major glucose transporter, and therefore enhance uptake of vitamin C via its oxidized form dehydroascorbate (DHA), and then reduce DHA inside the cells at the expense of glutathione-induced oxidative stress. This hypothesis emphasizes a role of DHA in inducing intracellular oxidative stress. However, data from other investigators do not support this hypothesis. First, if DHA uptake is mediating the oxidative stress, then treatment with DHA would induce cancer cell death. But it did not, and DHA did not reduce GAPDH activity. In fact, DHA does not exhibit any of the effects ascorbate has on cancer cells [1,13,36,37]. Pancreatic cancer cells with wild-type *KRAS* showed no difference in sensitivity to ascorbate than those harboring *KRAS* mutations [32]. Blocking GLUT1 transporters using 2-deoxy-D-glucose failed to rescue ascorbate-mediated cytotoxicity in non-small cell lung cancer cell lines [9]. In fact, an animal study using DHA found it counteracts with chemotherapy, in contrast to the vast majority of studies using ascorbate [37].

More mechanistic details have been revealed after ATP drop in ascorbate-treated cancer cells. AMP-activated protein kinase (AMPK) is the main sensor of cellular energy status and is activated when cellular AMP:ATP and ADP:ATP ratio is increased. AMPK activation restores the energy balance by inhibiting the anabolic process, which utilizes ATP (protein synthesis, fatty acid synthesis, sterol synthesis), while promoting catabolic processes that generate ATP (fatty acid oxidation, glucose uptake, glycolysis, autophagy) [38]. Ascorbate treatment resulted in activation of AMPK as evidenced by increased phosphorylation of the catalytic subunit of AMPK, AMPK-α. Activation of AMPK in cancer cells led to the inhibition of protein synthesis and cell growth by inhibiting mTOR (Figure 3.2), which is a central regulator of many biosynthetic pathways, especially protein translation [4]. Currently, it has been proposed, but efforts have not been made to discover the other signaling networks altered by AMPK when cells are treated with pharmacologic ascorbate.

INTERRUPTION IN MICROTUBULE DYNAMICS

Pharmacologic concentrations of ascorbate have been shown to cause cell cycle arrest (G2/M phase blockade) [14], inhibition of cell growth and cell division [4,8,14,39], and inhibition of cancer cell metastasis [8]. The cell signaling mechanism that regulates these processes is not fully understood. Polireddy et al. identified a novel mechanism through which ascorbate regulated cell cycle arrest and inhibition of metastasis. These studies revealed a connection between ascorbate and microtubule dynamics.

Microtubules consist of α-tubulin and β-tubulin heterodimer, assembled end to end into a long filament called **protofilament**. Approximately 13 protofilaments laterally associate to form a hollow microtubule cylinder. Microtubules are highly dynamic. They constantly switch between polymerized and depolymerized states, leading to the generation of highly specialized structures such as mitotic spindles during cell division. They also act as tracks for the transport of cargo molecules within the cells and are important

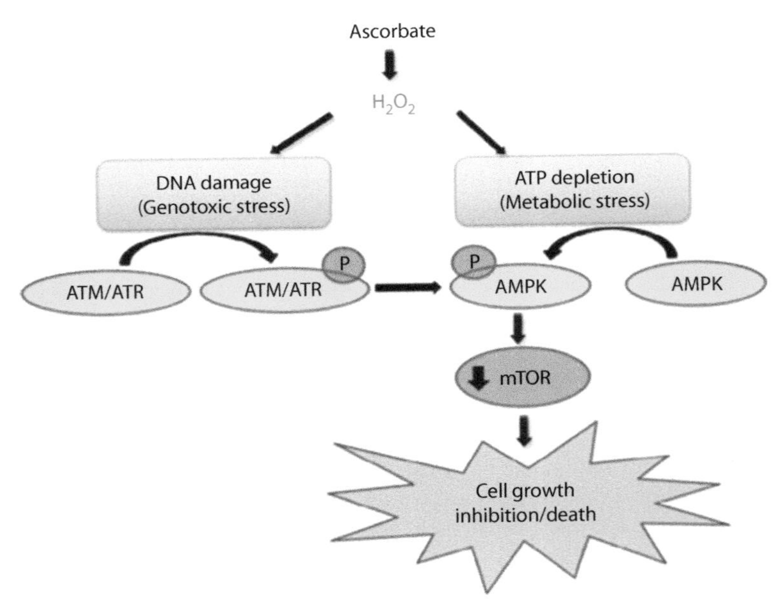

Figure 3.2. Downstream mechanisms of ATP depletion induced by pharmacologic ascorbate. The DNA damage and ATP depletion cause phosphorylation of the downstream sensors: The DNA damage sensor ATM/ATR, and the cellular energy sensor AMPK. Activation of AMPK leads to inhibition in mTOR, a central regulator for protein synthesis, and consequently results in cell growth inhibition, or cell death. (AMPK, 5'-adenosine monophosphate-activated protein kinase; ATR, ATM- and Rad3-related kinase.)

for the mobility of the cells. Thus, regulation of microtubule dynamics dictates its functions [40].

Acetylation of Lys-40 on an α-tubulin subunit stabilizes lateral interactions between neighboring tubulin protofilaments, whereas deacetylation enhances tubulin depolymerization [41]. Therefore, the acetylation status of α-tubulin confers functional diversity of tubulin polymers by altering its dynamics. In pancreatic cancer cells (PANC-1 and BxPC-3), ascorbate at both subcytotoxic and cytotoxic concentrations (1.25–2.5 mM) increased tubulin acetylation in a dose-dependent manner. Stable microtubules were generated and failed to undergo depolymerization, even in the presence of a tubulin depolymerizing agent, nocodazole [8]. Interestingly, the increase of α-tubulin acetylation in a normal pancreatic ductal epithelial cell was minimal. H_2O_2 treatment mimicked the effects of ascorbate, and catalase completely prevents the effects of ascorbate. These data indicate that ascorbate stabilizes α-tubulin, and the action is dependent on H_2O_2 [8] (Figure 3.3). The same study also reported that ascorbate treatment increased expression of epithelial markers (E-cadherin) in pancreatic cancer cells and decreased mesenchymal markers (N-Cad and vimentin), which ultimately increases cell-cell contact and reduces cancer cell metastasis [8]. Polireddy's study also found decreased metalloproteinases (MMPs) and increased collagen in pancreatic cancers treated with ascorbate. Whether the changes in the epithelial-mesenchymal transition (EMT) and cell-cell contact are mechanistically independent or interrelated to tubulin dynamics is not known. A similar phenomenon was observed in a study conducted in endothelial cells [42]. It was found that when ascorbate stabilized microtubules and increased acetylation of α-tubulin, it also increased cell-cell contact and tightened the endothelial permeability barrier [42]. Nevertheless, as a result, the migration and invasion ability of these pancreatic cancer cells decreased, as well as the proliferation [8].

Tubulin acetylation is under the control of balanced enzymatic activities of acetyltransferases (α-tubulin acetyltransferase [α-TAT]) and deacetylases (histone deacetylase 6 [HDAC6] and Sirtuin 2 [Sirt-2]). Sirt-2 is a NAD^+-dependent protein deacetylase, whereas HDAC6 is a Zn^{+2} dependent protein deacetylase. Both Sirt-2 and HDAC6 predominantly localize to the cytoplasm and have deacetylase activity for αK40 [43–45]. Ascorbate has no effect on the expression levels of α-TAT and Sirt-2 but slightly decreased

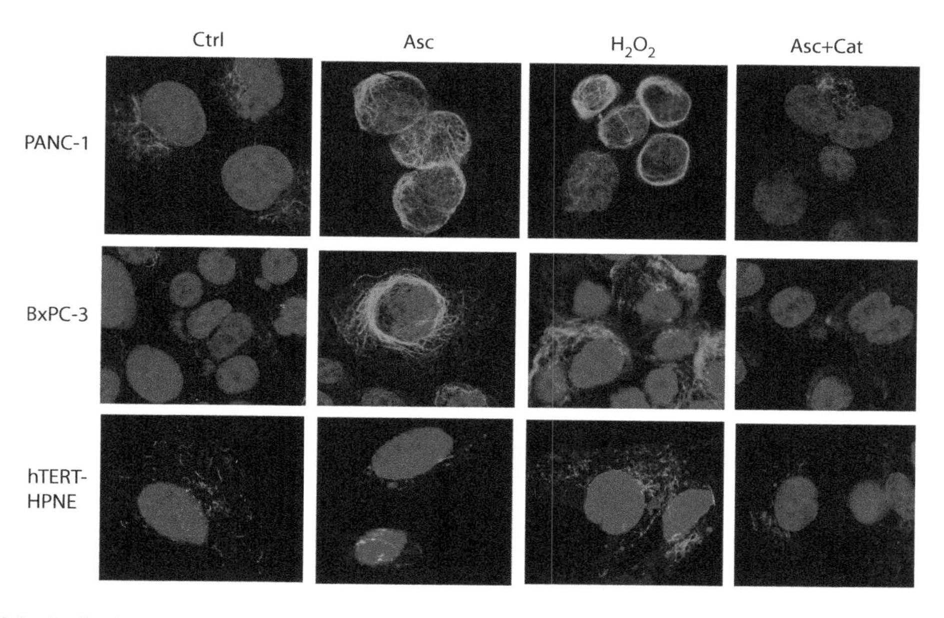

Figure 3.3. Confocal microscopy showing immunofluorescence of tubulin acetylation in pancreatic cancer cells treated with pharmacologic ascorbate. Green shows acetylated α-tubulin stained with a fluorocore-conjugated antiacetylated α-tubulin antibody. Blue shows cell nuclei stained with hoechst33342. PANC-1 and BxPC-3 were human pancreatic cancer cells, and hTERT-HPNE was a human normal pancreatic ductal epithelial cell line. Cells were treated for 4 hours with ascorbate (Asc, 2.5 mM) or hydrogen peroxide (H_2O_2, 500 μM) or Asc + Cat (catalase, 600 U/mL).

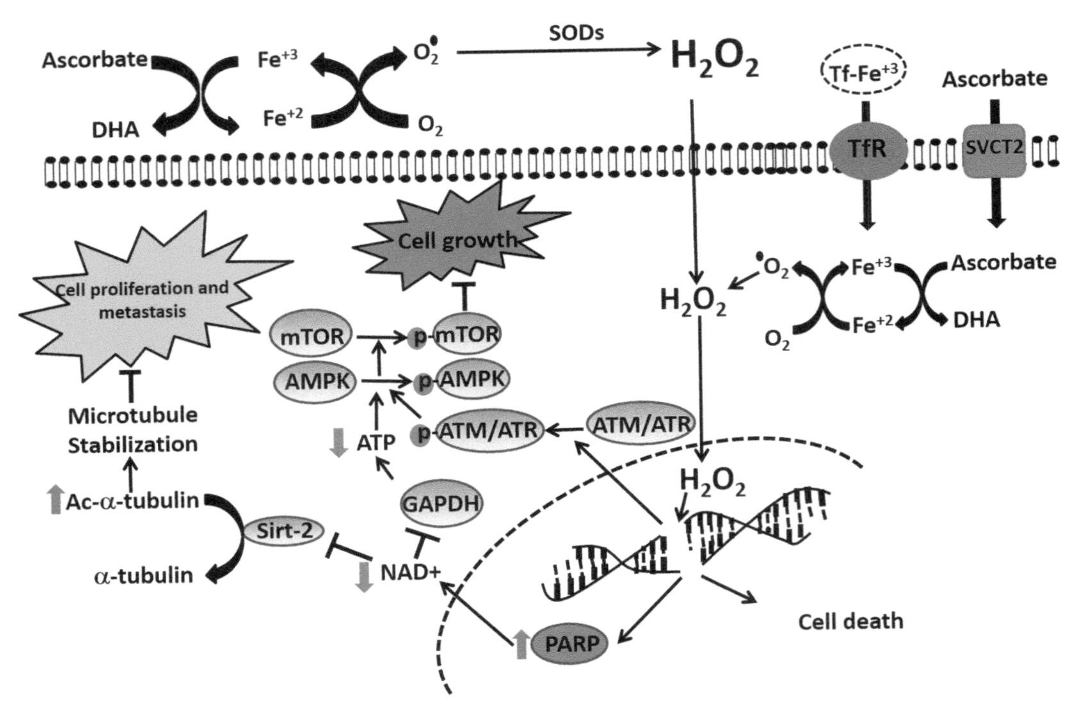

Figure 3.4. A summary of mechanisms of ascorbate-induced cell death discussed in this chapter. (AMPK, 5'-adenosine monophosphate-activated protein kinase; ATM, ataxia-telangiectasia mutated; ATR, ATM- and Rad3-related kinase; DHA, dehydroascorbate; GAPDH, glyceraldehyde 3-phosphate dehydrogenase; mTOR, mammalian target of rapamycin; PARP, poly-ADP-ribose polymerase; SOD, superoxide dismutase; SVCT2, sodium-dependent vitamin C transporter 2; TfR, transferrin receptor.)

HDAC6 expression. However, overexpression of HDAC6 in pancreatic cancer cell lines failed to rescue ascorbate-mediated α-tubulin acetylation, suggesting that HDAC6 has a minimal role in this process despite its expression being influenced by ascorbate. While the influence of ascorbate on α-TAT is less studied, the current studies demonstrated the importance of Sirt-2. Ascorbate enhances α-tubulin acetylation mainly by inhibiting Sirt-2 activity, as a consequence of NAD^+ depletion. Supplementation of NAD^+ prevented the ascorbate-mediated α-tubulin acetylation. Results from these studies are critical for defining the synergistic effect of ascorbate and tubulin-stabilizing anticancer drugs, such as paclitaxel [4].

DISCUSSION

In summary, understanding the anticancer activity of ascorbate provides a strong basis for its utility in the clinic for the treatment of cancer. Ascorbate elicited multiple signaling pathways exemplified in this chapter to inhibit cancer cell proliferation, metastasis, and cell death (Figure 3.4). We believe there are more pathways and signaling networks that are involved in causing selective cancer cell inhibition by pharmacologic ascorbate. Some of the other known possibilities include hypoxia-inducible factor (HIF) [46] and epigenetic modification (e.g., through Tet2) [47], which are discussed separately in other chapters of this book. These (and potentially more) pathways work in concert to exert the preferential inhibition that pharmacologic ascorbate has on cancer cells, and spare or even protect normal cells.

At first glance, the multitargeting feature of ascorbate is disadvantageous because promiscuous drug mechanisms of action usually raise the concern of toxicity. However, in the case of ascorbate, the toxic side effects prove to be minimal, if there are any, in multiple clinical trials, even with doses as high as 75–125 gram/infusion, two to three times per week for months, or less frequent but prolonged usage up to 1 year in cancer patients [4,48–58]. The multitargeting feature of ascorbate is in fact advantageous, because it would provide many opportunities for combination therapy aiming for better efficacy, and it would reduce the likelihood of resistance establishment in cancer cells. Cancer cells are heterogeneous even within one tumor, and they are prone to mutate under the pressure of treatment. Targeted therapies are often compromised and may ultimately fail because of mutations. This makes pharmacologic ascorbate a unique agent, because even if inhibition of a target is incomplete, or mutations render it ineffective in targeting a particular pathway, treatment effects could be exerted through other pathways, especially when combined with different anticancer therapies.

In spite of all the cellular pathways that ascorbate influences, it is apparent that peroxide formation is the essential mechanism for ascorbate-induced cancer cell death. There continue to be debates on whether uptake of ascorbate into the cells plays a role in the anticancer effects. By high-dose intravenous infusion of ascorbate, one can dramatically raise the ascorbate concentrations in the extracellular milieu, while the intracellular concentrations are relatively consistent when there is no ascorbate deficiency at the baseline. When ascorbate concentrations increase to millimolar range in the extracellular fluids, ascorbate radicals and H_2O_2 form, dependent on ascorbate concentration and time, but do not require the presence of cells [1,2]. Catalase, with a molecular weight of 232 kD [59], does not get into cells when added to the culture media, and the addition of catalase completely protects the cells from ascorbate-induced death. Cells with preloaded ascorbate respond the same to pharmacologic ascorbate treatment as their unloaded counterparts [1]. The evidence suggests that when the extracellular ascorbate concentrations rise to pharmacologic range, peroxide is generated and is essential and sufficient to induce death in sensitive cancer cells [1–3,32].

Note that cancer patients are often insufficient in ascorbate. There is the possibility that an increase in intracellular ascorbate concentration plays a role in cancer cell killing, for example, by enhancing intracellular peroxide formation [9], by modifying epigenetics [47,60], and by influencing HIF1α [61] in these cells. A recent study showed that high-dose ascorbate selectively inhibited hepatic cancer stem-like cells that overexpressed SVCT2, the sodium-dependent vitamin C transporter [11].

Besides the mechanisms that have been investigated, there are potential effects of high-dose ascorbate on other components of the cells that are less explored. Two of the less explored examples are effects on cytoplasmic membrane dynamics and cellular cargo transportation.

The cytoplasmic membrane is a target for H_2O_2. Studies showed that exogenous addition of H_2O_2 to media altered fluidity and leakiness of cultured pulmonary artery endothelial cell membrane, which can be prevented by pretreatment with vitamin E, indicating lipid peroxidation caused by H_2O_2 [62]. H_2O_2 is also known to cause dysfunction of membrane proteins through oxidation of sulfhydryl groups [63]. Researchers are still at the infancy stage in understanding the effects of pharmacologic ascorbate on membrane fluidity dynamics as well as functional alteration of membrane proteins. Interestingly, subtoxic (50 uM), short-term exposure of HL-60 cells and human skin fibroblasts to H_2O_2 has been shown to cause intracellular accumulation of daunorubicin, a chemotherapeutic drug most commonly used to treat acute myelogenous leukemia and acute lymphoblastic leukemia. Though the exact mechanism of drug accumulation is not known, it was suggested that H_2O_2 decreased lateral diffusion of the cytoplasmic membrane of the endothelial cells, ultimately leading to the accumulation of daunorubicin [64]. It will be interesting to see if ascorbate has any effect on the accumulation of chemotherapeutic drugs in cancer cells.

Microtubules act as tracks for cargo transport with the help of motor proteins (kinesin and dynein) in and out of the cells. The motor proteins kinesin and dynein associate with cargoes and transport them along microtubules. Tubulin posttranslational modifications are associated with the recruitment of specific types of motor molecules. For example, acetylated α-tubulin specifically interacts with kinesin 1 cargo complex, whereas tyrosinated α-tubulin interacts with kinesin 3 cargo complex [65]. Assuming from the fact that pharmacologic ascorbate enhanced α-tubulin acetylation, it is possible that cargo transport is influenced. Currently, there efforts have not been made to understand the effect of ascorbate on cargo transport mediated by motor proteins. This question is particularly important in neuronal transport, that ascorbate might have pathophysiologic or therapeutic implication for diseases of the nervous system. Microtubules in the axon organize into bundles and enable efficient transport of neurotransmitters. Such bundled microtubules are also observed in primary cilia and flagella, as well as in mitotic spindles. Often, these microtubules are marked by acetylation. Further research is required to address these intriguing questions.

REFERENCES

1. Chen, Q., Espey, M. G., Krishna, M. C., Mitchell, J. B., Corpe, C. P., Buettner, G. R., Shacter, E. and Levine, M. 2005 Pharmacologic ascorbic acid concentrations selectively kill cancer cells: Action as a pro-drug to deliver hydrogen peroxide to tissues. *Proc. Natl. Acad. Sci. USA* 102, 13604–13609.
2. Chen, Q., Espey, M. G., Sun, A. Y., Lee, J. H., Krishna, M. C., Shacter, E., Choyke, P. L. et al. 2007 Ascorbate in pharmacologic concentrations selectively generates ascorbate radical and hydrogen peroxide in extracellular fluid in vivo. *Proc. Natl. Acad. Sci. USA* 104, 8749–8754.
3. Chen, Q., Espey, M. G., Sun, A. Y., Pooput, C., Kirk, K. L., Krishna, M. C., Khosh, D. B., Drisko, J. and Levine, M. 2008 Pharmacologic doses of ascorbate act as a prooxidant and decrease growth of aggressive tumor xenografts in mice. *Proc. Natl. Acad. Sci. USA* 105, 11105–11109.
4. Ma, Y., Chapman, J., Levine, M., Polireddy, K., Drisko, J. and Chen, Q. 2014 High-dose parenteral ascorbate enhanced chemosensitivity of ovarian cancer and reduced toxicity of chemotherapy. *Sci. Transl. Med.* 6, 222ra218.
5. Yun, J., Mullarky, E., Lu, C., Bosch, K. N., Kavalier, A., Rivera, K., Roper, J. et al. 2015 Vitamin C selectively kills KRAS and BRAF mutant colorectal cancer cells by targeting GAPDH. *Science* 350, 1391–1396.
6. Schoenfeld, J. D., Sibenaller, Z. A., Mapuskar, K. A., Wagner, B. A., Cramer-Morales, K. L., Furqan, M., Sandhu, S. et al. 2017 O2- and H_2O_2-mediated disruption of Fe Metabolism causes the differential susceptibility of NSCLC and GBM cancer cells to pharmacological ascorbate. *Cancer Cell* 31, 487–500.
7. Chen, P., Stone, J., Sullivan, G., Drisko, J. A. and Chen, Q. 2011 Anti-cancer effect of pharmacologic ascorbate and its interaction with supplementary parenteral glutathione in preclinical cancer models. *Free Radic. Biol. Med.* 51, 681–687.
8. Polireddy, K., Dong, R., Reed, G., Yu, J., Chen, P., Williamson, S., Violet, P. C. et al. 2017 High dose parenteral ascorbate inhibited pancreatic cancer growth and metastasis: Mechanisms and a phase I/IIa study. *Sci. Rep.* 7(1), 17188.
9. Schoenfeld, J. D., Sibenaller, Z. A., Mapuskar, K. A., Wagner, B. A., Cramer-Morales, K. L., Furqan, M., Sandhu, S. et al. 2017 O2(-) and H_2O_2-mediated disruption of Fe Metabolism causes the differential susceptibility of NSCLC and GBM cancer cells to pharmacological ascorbate. *Cancer Cell* 31, 487–500 e488.
10. Hong, S. W., Lee, S. H., Moon, J. H., Hwang, J. J., Kim, D. E., Ko, E., Kim, H. S. et al. 2013 SVCT-2 in breast cancer acts as an indicator for L-ascorbate treatment. *Oncogene* 32, 1508–1517.
11. Lv, H., Wang, C., Fang, T., Li, T., Lv, G., Han, Q., Yang, W. and Wang, H. 2018 Vitamin C preferentially kills cancer stem cells in hepatocellular carcinoma via SVCT-2. *NPJ. Precis. Oncol.* 2, 1.
12. Berquist, B. R., Wilson, D. M., 3rd. 2012 Pathways for repairing and tolerating the spectrum of oxidative DNA lesions. *Cancer Lett.* 327, 61–72.
13. Ma, E., Chen, P., Wilkins, H. M., Wang, T., Swerdlow, R. H. and Chen, Q. 2017 Pharmacologic ascorbate induces neuroblastoma cell death by hydrogen peroxide mediated DNA damage and reduction in cancer cell glycolysis. *Free Radic. Biol. Med.* 113, 36–47.
14. Herst, P. M., Broadley, K. W., Harper, J. L. and McConnell, M. J. 2012 Pharmacological concentrations of ascorbate radiosensitize glioblastoma multiforme primary cells by increasing oxidative DNA damage and inhibiting G2/M arrest. *Free Radic. Biol. Med.* 52, 1486–1493.
15. Kaina, B., Fritz, G. 2006 *DNA Damaging Agents, Encyclopedic Reference of Genomics and Proteomics in Molecular Medicine*. Springer Berlin Heidelberg, Berlin, Heidelberg, pp. 416–423.
16. Zhou, B. B. and Elledge, S. J. 2000 The DNA damage response: Putting checkpoints in perspective. *Nature* 408, 433–439.
17. Kim, J., Lee, S. D., Chang, B., Jin, D. H., Jung, S. I., Park, M. Y., Han, Y. et al. 2012 Enhanced antitumor activity of vitamin C via p53 in cancer cells. *Free Radic. Biol. Med.* 53, 1607–1615.
18. Lavin, M. F. 2008 Ataxia-telangiectasia: From a rare disorder to a paradigm for cell signalling and cancer. *Nat. Rev. Mol. Cell Biol.* 9, 759–769.
19. Shiloh, Y. 2003 ATM and related protein kinases: Safeguarding genome integrity. *Nat. Rev. Cancer* 3, 155–168.
20. Ceccaldi, R., Rondinelli, B. and D'Andrea, A. D. 2016 Repair pathway choices and consequences at the double-strand break. *Trends Cell Biol.* 26, 52–64.

21. Davis, A. J. and Chen, D. J. 2013 DNA double strand break repair via non-homologous end-joining. *Transl. Cancer. Res.* 2, 130–143.
22. Ma, Y., Chen, P., Drisko, J., Godwin, A., Chen, Q. 2019 Pharmacologic ascorbate induces "BRCAness" and enhances effects of poly(ADP-Ribose) polymerase inhibitors against BRCA1/2 wild-type ovarian cancer, oncology letters, in revision.
23. Schraufstatter, I. U., Hyslop, P. A., Hinshaw, D. B., Spragg, R. G., Sklar, L. A. and Cochrane, C. G. 1986 Hydrogen peroxide-induced injury of cells and its prevention by inhibitors of poly(ADP-ribose) polymerase. *Proc. Natl. Acad. Sci. USA* 83, 4908–4912.
24. Schraufstatter, I. U., Hinshaw, D. B., Hyslop, P. A., Spragg, R. G. and Cochrane, C. G. 1986 Oxidant injury of cells. DNA strand-breaks activate polyadenosine diphosphate-ribose polymerase and lead to depletion of nicotinamide adenine dinucleotide. *J. Clin. Invest.* 77, 1312–1320.
25. Warburg, O., Wind, F. and Negelein, E. 1927 The metabolism of tumors in the body. *J. Gen. Physiol.* 8, 519–530.
26. Liberti, M. V. and Locasale, J. W. 2016 The Warburg effect: How does it benefit cancer cells? *Trends Biochem. Sci.* 41, 211–218.
27. Chen, P., Yu, J., Chalmers, B., Drisko, J., Yang, J., Li, B. and Chen, Q. 2012 Pharmacological ascorbate induces cytotoxicity in prostate cancer cells through ATP depletion and induction of autophagy. *Anticancer Drugs* 23, 437–444.
28. Verrax, J., Dejeans, N., Sid, B., Glorieux, C. and Calderon, P. B. 2011 Intracellular ATP levels determine cell death fate of cancer cells exposed to both standard and redox chemotherapeutic agents. *Biochem. Pharmacol.* 82, 1540–1548.
29. Verrax, J., Vanbever, S., Stockis, J., Taper, H. and Calderon, P. B. 2007 Role of glycolysis inhibition and poly(ADP-ribose) polymerase activation in necrotic-like cell death caused by ascorbate/menadione-induced oxidative stress in K562 human chronic myelogenous leukemic cells. *Int. J. Cancer* 120, 1192–1197.
30. Lee, Y. J. and Shacter, E. 1999 Oxidative stress inhibits apoptosis in human lymphoma cells. *J. Biol. Chem.* 274, 19792–19798.
31. Filipovic, D. M., Meng, X. and Reeves, W. B. 1999 Inhibition of PARP prevents oxidant-induced necrosis but not apoptosis in LLC-PK1 cells. *Am. J. Physiol.* 277, F428–436.
32. Du, J., Martin, S. M., Levine, M., Wagner, B. A., Buettner, G. R., Wang, S. H., Taghiyev, A. F., Du, C., Knudson, C. M. and Cullen, J. J. 2010 Mechanisms of ascorbate-induced cytotoxicity in pancreatic cancer. *Clin. Cancer. Res.* 16, 509–520.
33. Ostrem, J. M. and Shokat, K. M. 2016 Direct small-molecule inhibitors of KRAS: From structural insights to mechanism-based design. *Nat. Rev. Drug. Discov.* 15, 771–785.
34. Gupta, A. K., Wang, X., Pagba, C. V., Prakash, P., Sarkar-Banerjee, S., Putkey, J. and Gorfe, A. A. 2019 Multi-target, ensemble-based virtual screening yields novel allosteric KRAS inhibitors at high success rate. *Chem. Biol. Drug. Des.* 94, 1441–1456.
35. McCarthy, M. J., Pagba, C. V., Prakash, P., Naji, A. K., van der Hoeven, D., Liang, H., Gupta, A. K. et al. 2019 Discovery of high-affinity noncovalent allosteric KRAS inhibitors that disrupt effector binding. *ACS Omega* 4, 2921–2930.
36. Fischer, A. P. and Miles, S. L. 2016 Ascorbic acid, but not dehydroascorbic acid increases intracellular vitamin C content to decrease hypoxia inducible factor-1 alpha activity and reduce malignant potential in human melanoma. *Biomed. Pharmacother.* 86, 502–513.
37. Heaney, M. L., Gardner, J. R., Karasavvas, N., Golde, D. W., Scheinberg, D. A., Smith, E. A. and O'Connor, O. A. 2008 Vitamin C antagonizes the cytotoxic effects of antineoplastic drugs. *Cancer Res.* 68, 8031–8038.
38. Garcia, D. and Shaw, R. J. 2017 AMPK: Mechanisms of cellular energy sensing and restoration of metabolic balance. *Mol. Cell* 66, 789–800.
39. Maramag, C., Menon, M., Balaji, K. C., Reddy, P. G. and Laxmanan, S. 1997 Effect of vitamin C on prostate cancer cells in vitro: Effect on cell number, viability, and DNA synthesis. *Prostate* 32, 188–195.
40. Brouhard, G. J. and Rice, L. M. 2018 Microtubule dynamics: An interplay of biochemistry and mechanics. *Nat. Rev. Mol. Cell Biol.* 19, 451–463.
41. Cueva, J. G., Hsin, J., Huang, K. C. and Goodman, M. B. 2012 Posttranslational acetylation of alpha-tubulin constrains protofilament number in native microtubules. *Curr. Biol.* 22, 1066–1074.
42. Parker, W. H., Rhea, E. M., Qu, Z. C., Hecker, M. R. and May, J. M. 2016 Intracellular ascorbate tightens the endothelial permeability barrier through Epac1 and the tubulin cytoskeleton. *Am. J. Physiol. Cell Physiol.* 311, C652–C662.
43. Li, L. and Yang, X. J. 2015 Tubulin acetylation: Responsible enzymes, biological functions and human diseases. *Cell Mol. Life Sci.* 72, 4237–4255.
44. Hubbert, C., Guardiola, A., Shao, R., Kawaguchi, Y., Ito, A., Nixon, A., Yoshida, M., Wang, X. F. and Yao, T. P. 2002 HDAC6 is a microtubule-associated deacetylase. *Nature* 417, 455–458.

45. North, B. J., Marshall, B. L., Borra, M. T., Denu, J. M. and Verdin, E. 2003 The human Sir2 ortholog, SIRT2, is an NAD+-dependent tubulin deacetylase. *Mol. Cell* 11, 437–444.
46. Wilkes, J. G., O'Leary, B. R., Du, J., Klinger, A. R., Sibenaller, Z. A., Doskey, C. M., Gibson-Corley, K. N. et al. 2018 Pharmacologic ascorbate (P-AscH(-)) suppresses hypoxia-inducible factor-1alpha (HIF-1alpha) in pancreatic adenocarcinoma. *Clin. Exp. Metastasis.* 35, 37–51.
47. Agathocleous, M., Meacham, C. E., Burgess, R. J., Piskounova, E., Zhao, Z., Crane, G. M., Cowin, B. L. et al. 2017 Ascorbate regulates haematopoietic stem cell function and leukaemogenesis. *Nature* 549, 476–481.
48. Stephenson, C. M., Levin, R. D., Spector, T. and Lis, C. G. 2013 Phase I clinical trial to evaluate the safety, tolerability, and pharmacokinetics of high-dose intravenous ascorbic acid in patients with advanced cancer. *Cancer Chemother. Pharmacol.* 72, 139–146.
49. Welsh, J. L., Wagner, B. A., van't Erve, T. J., Zehr, P. S., Berg, D. J., Halfdanarson, T. R., Yee, N. S. et al. 2013 Pharmacological ascorbate with gemcitabine for the control of metastatic and node-positive pancreatic cancer (PACMAN): Results from a phase I clinical trial. *Cancer Chemother. Pharmacol.* 71, 765–775.
50. Monti, D. A., Mitchell, E., Bazzan, A. J., Littman, S., Zabrecky, G., Yeo, C. J., Pillai, M. V., Newberg, A. B., Deshmukh, S. and Levine, M. 2012 Phase I evaluation of intravenous ascorbic acid in combination with gemcitabine and erlotinib in patients with metastatic pancreatic cancer. *PLOS ONE* 7, e29794.
51. Mikirova, N., Casciari, J., Rogers, A. and Taylor, P. 2012 Effect of high-dose intravenous vitamin C on inflammation in cancer patients. *J. Transl. Med.* 10, 189.
52. Vollbracht, C., Schneider, B., Leendert, V., Weiss, G., Auerbach, L. and Beuth, J. 2011 Intravenous vitamin C administration improves quality of life in breast cancer patients during chemo-/radiotherapy and aftercare: Results of a retrospective, multicentre, epidemiological cohort study in Germany. *In Vivo* 25, 983–990.
53. Hoffer, L. J., Levine, M., Assouline, S., Melnychuk, D., Padayatty, S. J., Rosadiuk, K., Rousseau, C., Robitaille, L., Miller, W.H., Jr. 2008 Phase I clinical trial of i.v. ascorbic acid in advanced malignancy. *Ann. Oncol.* 19, 1969–1974.
54. Yeom, C. H., Jung, G. C. and Song, K. J. 2007 Changes of terminal cancer patients' health-related quality of life after high dose vitamin C administration. *J. Korean Med. Sci.* 22, 7–11.
55. Riordan, H. D., Casciari, J. J., Gonzalez, M. J., Riordan, N. H., Miranda-Massari, J. R., Taylor, P. and Jackson, J. A. 2005 A pilot clinical study of continuous intravenous ascorbate in terminal cancer patients. *P. R. Health Sci. J.* 24, 269–276.
56. Cameron, E. and Campbell, A. 1991 Innovation vs. quality control: An "unpublishable" clinical trial of supplemental ascorbate in incurable cancer. *Med. Hypotheses.* 36, 185–189.
57. Cameron, E. and Pauling, L. 1976 Supplemental ascorbate in the supportive treatment of cancer: Prolongation of survival times in terminal human cancer. *Proc. Natl. Acad. Sci. USA* 73, 3685–3689.
58. Cameron, E. and Pauling, L. 1978 Supplemental ascorbate in the supportive treatment of cancer: Reevaluation of prolongation of survival times in terminal human cancer. *Proc. Natl. Acad. Sci. USA* 75, 4538–4542.
59. Schroeder, W. A., Shelton, J. R., Shelton, J. B., Robberson, B. and Apell, G. 1969 The amino acid sequence of bovine liver catalase: A preliminary report. *Arch. Biochem. Biophys.* 131, 653–655.
60. Minor, E. A., Court, B. L., Young, J. I. and Wang, G. 2013 Ascorbate induces ten-eleven translocation (Tet) methylcytosine dioxygenase-mediated generation of 5-hydroxymethylcytosine. *J. Biol. Chem.* 288, 13669–13674.
61. Vissers, M. C. M. and Das, A. B. 2018 Potential mechanisms of action for vitamin C in cancer: Reviewing the evidence. *Front. Physiol.* 9, 809.
62. Block, E. R. 1991 Hydrogen peroxide alters the physical state and function of the plasma membrane of pulmonary artery endothelial cells. *J. Cell Physiol.* 146, 362–369.
63. Snyder, L. M., Fortier, N. L., Leb, L., McKenney, J., Trainor, J., Sheerin, H. and Mohandas, N. 1988 The role of membrane protein sulfhydryl groups in hydrogen peroxide-mediated membrane damage in human erythrocytes. *Biochim. Biophys. Acta.* 937, 229–240.
64. Funk, R. S. and Krise, J. P. 2007 Exposure of cells to hydrogen peroxide can increase the intracellular accumulation of drugs. *Mol. Pharm.* 4, 154–159.
65. Ross, J. L., Ali, M. Y. and Warshaw, D. M. 2008 Cargo transport: Molecular motors navigate a complex cytoskeleton. *Curr. Opin. Cell Biol.* 20, 41–47.

CHAPTER FOUR

Ascorbate and the Hypoxic Response in Cancer

Christina Wohlrab, Caroline Kuiper, and Gabi U. Dachs

CONTENTS

Vitamin C (ascorbate) is an essential micronutrient for humans and must be acquired via our diet. Humans are unable to synthesize ascorbate due to mutations in gulonolactone oxidase (GULO), the terminal enzyme of the biosynthesis pathway [1]. Although our absolute requirement for this vitamin has long been established, its many roles in health and disease, and its mechanisms of action, are only now being uncovered. Here we describe how ascorbate may affect cancer progression via modulation of the hypoxic response pathway.

TUMOR HYPOXIA AND THE MICROENVIRONMENT

The growth of malignant cells requires the acquisition of several characteristics, allowing them to proliferate and thrive. These proposed "hallmarks of cancer" include sustained proliferative signaling, resistance to cell death, evasion of growth suppressors, invasion and metastasis, replicative immortality, and angiogenesis [2]. These features allow the formation of a tumor mass that is a heterogeneous and complex organ. Multiple cell types coexist in a tumor, with both cancerous and normal tissues dynamically interacting to create the tumor microenvironment, which is now recognized as a significant enabling factor in tumor growth and spread [2]. Conditions within the microenvironment are often inhospitable (reduced oxygenation, increased interstitial fluid pressure, low extracellular pH) (Figure 4.1) and thus highly conducive for Darwinian selection pressures to play out, which ultimately furthers tumor progression and spread [3].

One of the most well-characterized features of the tumor microenvironment is the reduction in available oxygen (hypoxia). The primary cause of tumor hypoxia is the highly disorganized and abnormal tumor microvasculature, comprising dilated, tortuous, and leaky vessels with disturbed arteriovenous flow dynamics [4]. The calculated diffusion distance of O_2 ranges from 41 to 183 μm [5], whereas the median intervessel distance in tumors is reported at ~257 μm, making spatial, or chronic, hypoxia a common feature of solid tumors [4]. At a distance beyond the oxygen diffusion limit, tumor cells become hypoxic

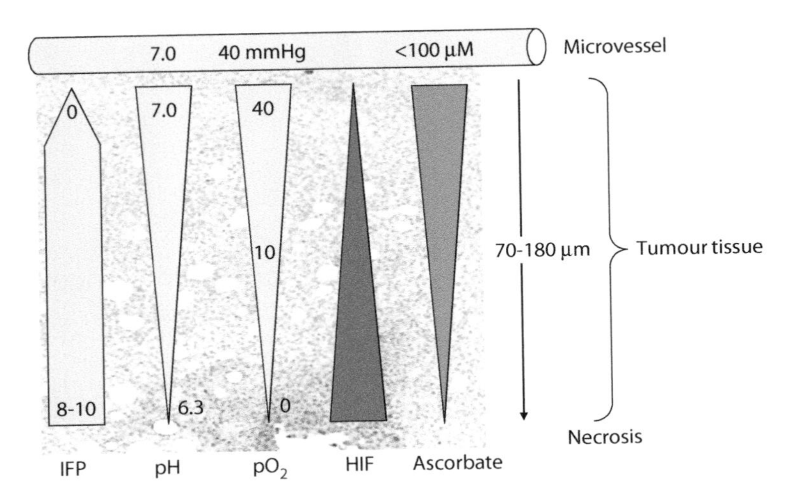

Figure 4.1. Relationship of microenvironmental factors, HIF, and ascorbate in a solid tumor mass. With increasing distance from blood vessels, extracellular pH drops from pH 7.0 and becomes acidic (pH 6.3), oxygen levels reduce from an average of 40 mm Hg and become hypoxic (10 mm Hg) or even anoxic (0 mm Hg), and ascorbate levels reduce (intracellular ascorbate levels range from about 0.3 nmol/μg DNA or 30 mg/100 g tissue to undetectable). With increasing distance from the blood supply, interstitial pressure rises, and protein levels and activity of the HIF transcription factor increase. (HIF, hypoxia inducible factor; IFP, interstitial fluid pressure [mm Hg]; pO_2, oxygen partial pressure [mm Hg].) Information collated from [2,3,10,11].

or even anoxic, with extended periods under these severe conditions leading to cell death and tissue necrosis (Figure 4.1). In addition, temporal fluctuations in pO_2 >5 mm Hg, primarily caused by changes in erythrocyte flux [6], can affect up to 50% of the tumor microvasculature and can range from seconds to minutes [7,8]. Hypoxia therefore occurs heterogeneously throughout the tumor mass, with both chronic and acute hypoxia affecting tumor cells, with its consequential biological effects still being investigated [9].

Hypoxia is associated with fundamental changes in cellular metabolism resulting in increased utilization of glycolysis (the Warburg effect) for energy generation and maintenance of vital biosynthesis pathways [10], leading to a reduction in extracellular pH (Figure 4.1). Reduced tumor blood flow is also associated with an increase in interstitial fluid pressure [11], which together result in poor delivery of therapeutic agents. Similar to oxygen, nutrients, and therapeutic agents, ascorbate will likely display a gradient in concentration with increased distance from the blood supply (Figure 4.1), although this has not yet been directly demonstrated *in vivo*.

HYPOXIA-INDUCIBLE FACTORS

Hypoxia-inducible factors (HIFs) respond to a reduction in cellular oxygenation by upregulating the production of a plethora of pro-survival factors [12,13]. HIFs are constitutively expressed heterodimeric transcription factors consisting of oxygen-sensitive α subunits (HIF-1α, HIF-2α, and HIF-3α) and a stable HIF-1β subunit, also known as aryl hydrocarbon receptor nuclear translocator (ARNT) [12,13] (Figure 4.2). Under activating conditions, the HIF-α/HIF-1β complex binds to hypoxia-regulated enhancers (HREs, consensus sequence 5′-RCGTG-3′) in gene promoters and recruits the CREB-binding protein (CBP)/p300 coactivator to induce target gene transcription [14,15]. HIF-1 and HIF-2 are well documented, whereas HIF-3 is relatively underexplored. HIF-3 reportedly has transcription factor activity, and a splice variant of HIF-3α is considered to act as a negative regulator of HIF-1 [16,17].

The HIF transcription factors regulate the expression of several hundred genes involved in metabolism, cell life and death pathways, glycolysis, and angiogenesis, and as a result, promote tumor growth, adaptation to the microenvironment, and resistance to chemo- and radiotherapy [15,18,19]. HIF-1α is present in all cells and tissues, whereas expression of HIF-2α is cell type specific and has a more limited distribution profile (endothelial cells, brain, pancreas, lungs, heart, intestines, tubular renal and hepatic cells) [20,21]. It is now generally accepted that despite structural similarities between HIF-1α and HIF-2α, these

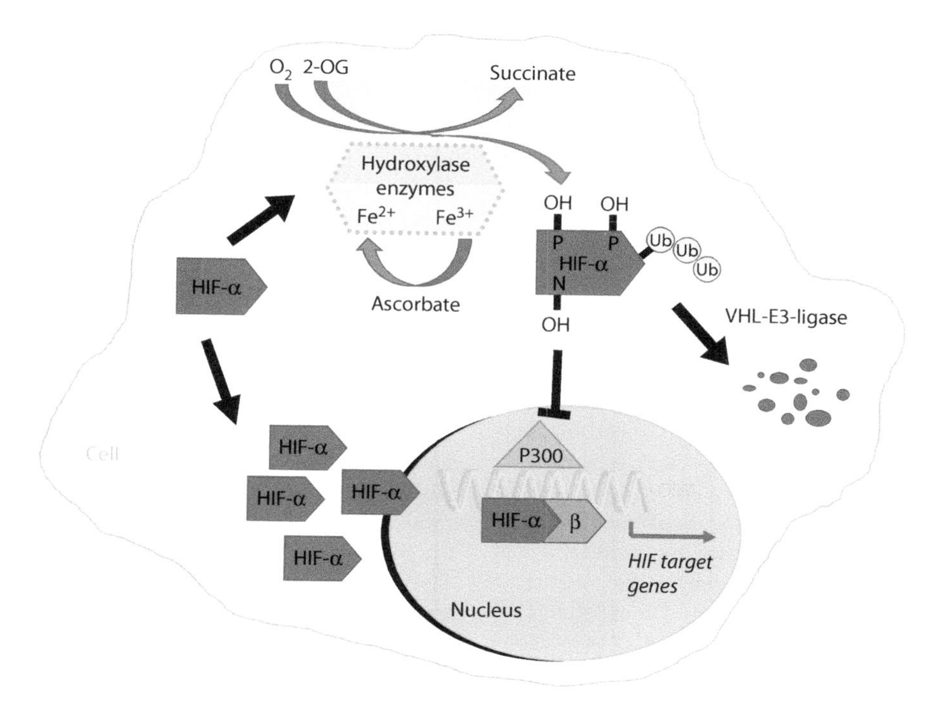

Figure 4.2. Regulation of HIF transcription factors by hydroxylase enzymes. The subunit HIF-α is constitutively expressed, and when the hydroxylase substrates O_2 and 2-OG, and/or cofactors Fe^{2-} and ascorbate are reduced, HIF-α accumulates, translocates to the nucleus where it binds to the HIF-β subunit, and with support of the p300 coactivator, form active transcription factors. When substrates and cofactors for the hydroxylase enzymes are supplied, hydroxylated HIF-α becomes a target for proteasomal degradation and becomes inactivated and unable to bind p300. (2-OG, 2-oxoglularate; HIF, hypoxia-inducible factor; OH, hydroxy; Ub, ubiquitin; VHL, von Hippel-Lindau.) Information collated from [13,18,59].

isoforms regulate the expression of common as well as unique target genes and therefore can have different or even opposing effects on cell fate [20,21]. Both HIF-1 and HIF-2 control angiogenesis via upregulation of the vascular endothelial growth factor (VEGF)-A and glucose uptake via the glucose transporter 1 (GLUT1); HIF-1 controls production of enzymes in the glycolytic pathway (including glyceraldehyde-3-phosphate dehydrogenase and phosphoglycerate kinase 1), controls regulation of pH via the carbonic anhydrase 9 (CA9), and acts as antiapoptotic factor by targeting the Bcl2/adenovirus E1B 19 kDa interacting protein 3 (BNIP3); HIF-2 controls erythropoietin (EPO) and is involved in cell cycle regulation via cyclin D1 and matrix remodeling via matrix metalloproteinase 1 [20–22]. Pan-genomic distribution of HIF binding sites, determined via chromatin immunoprecipitation coupled to next-generation DNA sequencing, demonstrated discrete binding patterns for HIF-1 and HIF-2 despite their identical consensus HRE sequence [23]. HIF-1 tended to bind within promoter regions, whereas HIF-2 bound more distal, and the two isoforms did not compete for binding sites [23].

HYPOXIA-INDUCIBLE FACTOR PATHWAY IN CANCER

HIF-1α protein is overexpressed in many cancers compared to preneoplastic lesions or normal tissue [24], and HIF-1α protein levels have been described as an independent prognostic indicator for poor outcome in many cancer studies. Meta-analyses have demonstrated an association between increased HIF-1α levels and lower patient survival in oral squamous cell carcinoma [25], small cell and non-small cell lung cancer [26], osteosarcoma [27], head and neck squamous cell carcinoma [28], pancreatic cancer [29], cervical cancer [30], epithelial ovarian cancer [31], hepatocellular carcinoma [32], esophageal cancer [33], breast cancer [34], and colorectal cancer [35], among others.

HIF-1 has been implicated in the progression from benign to preneoplastic to malignant lesions. For example, HIF-1α and specific target genes were increasingly expressed in human endometrial tissue, from normal to premalignant to adenocarcinoma [36] and were also increasingly expressed with each clinical stage of endometrial adenocarcinoma [37]. Similarly, HIF-1α protein was increasingly expressed in human gastric tissue, from gastritis to metaplasia to dysplasia to adenocarcinoma [38].

HIFs have also been implicated in cancer stem cell maintenance and are overexpressed in these cells even under normoxia [39]. Cancer stem cells represent a very small subset of cells within the tumor mass yet are believed to be responsible for initiating tumor growth and recurrence following therapy due to their high capacity for self-renewal.

HIF-regulated target genes are intimately involved in response to therapy. For example, HIF-1-dependent BNIP3 upregulation is responsible for hypoxia-induced resistance to etoposide treatment *in vitro*, via mitochondrial enlargement conferring protection against apoptosis [40]. Other mechanisms by which HIF-1 mediates hypoxia-induced chemoresistance include drug efflux, inhibition of cell senescence, inhibition of DNA damage, and decreased mitochondrial activity [41–44]. HIF-1 is also able to inhibit the oncogene MYC, resulting in cell cycle arrest and inhibition of DNA repair genes, which in turn will enhance response to chemotherapy [45], thus highlighting the complexity of the HIF-1 response.

Tumor hypoxia is clearly associated with resistance to radiotherapy due to the lack of oxygen-derived free radicals that cause DNA strand breaks on exposure to ionizing radiation [9,46]. However, HIF-1 has been associated with radioresistance independent of tumor oxygenation in xenograft studies [47]. In addition, siRNA inhibition of HIF-1α in prostate cancer cells significantly enhanced radiosensitization [48], while HIF-1 induction conferred radioresistance in cervical carcinoma cells [49]. However, basal levels of HIF-1 in several cancer cell lines did not correlate with radioresponsiveness [50].

One of the main mechanisms by which HIF-1 promotes resistance to both radiation and chemotherapy is the activation of DNA repair pathways [51]. Indeed, HIF-1 controls expression of most of the cell's DNA repair genes, including poly-ADP-ribose polymerase (PARP-1), XPA (part of base excision repair), ataxia telangiectasia mutated (ATM), and DNA-dependent protein kinase (DNA-PK) [52,53]. In addition, resistance to cancer therapies are induced via HIF-1-regulated metabolic reprogramming and modulation of cell death pathways, namely, inhibition of apoptosis and activation of autophagy [51].

REGULATION OF HYPOXIA-INDUCIBLE FACTORS

All HIF subunits have a basic helix-loop-helix domain and two per-arnt-sim (PAS) domains for dimerization and DNA binding, and a C-terminal transcriptional activation domain (CAD) [54]. The HIF-α subunits additionally have an N-terminal transcriptional activation domain (NAD). Both the CAD and NAD domains are involved in DNA binding, recruitment of coactivators, and transactivation activity of HIF [55]. The NAD is likely responsible for targeted gene specificity of HIF-1 and HIF-2 [56]. The N- and C-terminal oxygen-dependent degradation domains (NODD and CODD) of HIF-1α and HIF-2α enable regulation of protein stability [57].

HIFs are predominantly regulated at the post-translational (protein) level, although transcriptional and translational regulation of HIF levels has been reported [58]. Protein levels and activity of HIF-1 and HIF-2 are controlled via hydroxylation reactions by enzymes belonging to the family of iron- and 2-oxoglutarate-dependent dioxygenases [59] (Figure 4.2). Hydroxylation of specific prolines allows recognition of HIF-α by the von Hippel-Lindau (VHL) complex and leads to subsequent proteasomal degradation, and hydroxylation of a specific asparagine prohibits binding of transcriptional coactivators (CBP/p300) (Figure 4.2).

The dioxygenase enzymes have an absolute substrate requirement for oxygen and are therefore capable of oxygen sensing [60,61]. Inadequate supply with either of the two substrates, molecular oxygen and 2-oxoglutarate, or limited availability of the cofactors, iron and ascorbate, impairs activity of the enzymes. Hence, under hypoxia, hydroxylation of the HIF-α subunit cannot occur, and consequently, the interaction with the VHL complex and CBP/p300 is impaired, leading to stabilization and activation of HIF [59].

The proline hydroxylases (PHD1–3) target two prolines in the oxygen-dependent degradation domains of the HIF-α subunit (P402 and P564 of HIF-1α; P405 and P531 of HIF-2α) [57,62]. Hydroxylation of these amino acids enables binding of the VHL-E3 ubiquitin ligase complex and targets it for polyubiquitylation and proteasomal degradation (Figure 4.2). PHD2 is the dominant enzyme for hydroxylation of HIF-1α, and PHD1 and PHD3 have more influence on HIF-2α [63]. Hydroxylation of an asparagine in the CAD (N803 of HIF-1α, N847 of HIF-2α) by factor inhibiting HIF (FIH) disrupts recruitment of the coactivator CBP/p300, which interferes with the transactivation activity of HIF [60]. FIH has a higher affinity for oxygen reflected by a lower Km value (90 μM) than PHDs (230–250 μM) and remains active at mild hypoxia, which means it can reduce the transcriptional activity of HIF even in conditions where HIF protein is stabilized [64–66]. Of the two HIF isoforms, HIF-1α has been shown to be more sensitive than HIF-2α to regulation by FIH [67].

VHL is the substrate recognition unit of the E3 ligase complex that targets the HIF-α subunit for polyubiquitylation and proteasomal degradation under aerobic conditions [68]. VHL has a conserved hydroxyproline binding pocket in its β domain for substrate recognition and mediates binding to elongin C and cullin 2 via its α domain [69,70]. Dysfunction of the E3 ligase complex through mutations in *VHL* leads to loss of ability to bind hydroxylated HIF-α proteins, with resultant stabilization and accumulation of HIF proteins under normoxia. This triggers activation of genes that would normally be activated only under hypoxia [71,72]. VHL acts as a tumor suppressor, and mutations in *VHL* are associated with the inherited VHL cancer syndrome as well as sporadic cancers, including the predominant type of kidney cancer (clear cell renal cell carcinoma [ccRCC]), adrenal gland pheochromocytomas, and pancreatic neuroendocrine tumors [71].

HIF-HYDROXYLASES AND ASCORBATE

The structure of 2-oxoglutarate-dependent dioxygenases is described as a jelly-roll motif formed by two β-sheets consisting of eight antiparallel β-strands [73]. A nonheme iron (Fe[II]) is directly bound in the active site of the enzymes, coordinated by two histidines and one aspartate/glutamate. During the catalytic cycle, a further two iron coordination sites are occupied by 2-oxoglutarate and a final one by molecular oxygen [74].

In the hydroxylation reaction, the enzymes split molecular oxygen and transfer one atom to the target protein and the second to 2-oxoglutarate [75]. Binding of molecular oxygen to iron leads to decarboxylation of 2-oxoglutarate, resulting in the formation of succinate and CO_2, as well as the generation of a ferryl-oxo intermediate (Fe[IV]) which is highly reactive [76]. Fe(IV) is reduced to Fe(III) upon hydroxylation of the protein substrate [76]. Ascorbate is an established cofactor for this catalytic cycle [77] and has been proposed to be required for the regeneration of Fe(II) in the active center [78].

The requirement of the HIF hydroxylases for ascorbate is highlighted by their relatively high Km values (140–180 μM for PHDs, ~260 μM for FIH) [73]. The exact mechanism of ascorbate's role in supporting the hydroxylation reaction is still under investigation, but as ascorbate is a potent reducing agent, it has been proposed to act by converting Fe(III) back to Fe(II) at the end of the reaction cycle to keep the enzymes active [78]. More recent evidence has suggested that ascorbate might also function as a direct substrate for PHD enzymes, rather than only acting as a reducing agent [79]. Cell-free *in vitro* assays revealed that ascorbate acts on PHD in a saturating manner, and computer modeling of the enzyme crystal structure together with the vitamin showed accommodation of ascorbate in the active center of PHD2 [79]. A recent study has provided evidence that ascorbate supports the hydroxylation of full-length HIF-1α in live cells [80]. Other reducing agents (dithiothreitol, glutathione, and N-acetyl cysteine) are not able to replace ascorbate in the enzyme reaction [77,79]. FIH has been shown to be more susceptible to changes in ascorbate concentration, reflected by the higher Km values, compared to PHDs, and therefore, ascorbate might have a greater effect on transcriptional activity of HIF than on its protein levels [81].

ASSOCIATION OF ASCORBATE AND HYPOXIA-INDUCIBLE FACTORS IN CANCER CELLS

Although oxygenation plays a vital role, increasing intracellular ascorbate content is able to decrease protein levels of HIF-1α as well as reduce the expression of its downstream targets in cultured cancer cells [81–87]. Increasing concentrations of ascorbate were able to reduce hypoxia-induced HIF-1α under mild hypoxia (1% and 5% oxygen) but not under severe hypoxia (0.1% oxygen) in mouse lung cancer cells [88]. Treatment of human thyroid cancer cells with ascorbate *in vitro* resulted in a dose-dependent decrease of HIF-1α and GLUT-1 protein expression associated with increasing intracellular ascorbate levels [86].

The association of ascorbate and HIF-1α protein is supported by numerous studies, but the association of ascorbate with HIF-2α is underexplored and less clear. In thyroid cancer cells [86] and ccRCC cells [80], increasing ascorbate supply did not affect HIF-2α protein levels, whereas in colorectal cancer cells ascorbate increased HIF-2α protein [89].

DELIVERY AND UPTAKE OF ASCORBATE

Humans acquire ascorbate from their diet, where it is actively absorbed from the intestines primarily via sodium-dependent vitamin C transporter 1 (SVCT1), carried in the blood and taken up into all cells of the body, primarily via SVCT2, with a small fraction also taken up as dehydroascorbate via glucose transporters [90]. The SVCTs are capable of moving ascorbate against a concentration gradient, resulting in millimolar intracellular concentrations from micromolar plasma concentrations. Intracellular accumulation will vary depending on plasma availability, vascularization, tissue diffusion parameters, and cellular expression of ascorbate transporters [90].

Pharmacokinetic studies in healthy volunteers have demonstrated that dietary ascorbate results in a maximum plasma concentration of ~100 μM [91] due to controlled intestinal uptake and renal excretion of the excess [90]. This tight control can be bypassed by intravenous infusion of high doses of ascorbate, resulting in elevated plasma levels [91].

Cancer patients frequently have an inadequate (<50 μM plasma ascorbate) or even deficient (<11 μM) ascorbate status [92–94]. This is postulated to be due to reduced dietary intake, impaired uptake caused by surgery and treatment, or loss of ascorbate due to oxidation associated with inflammation and increased oxidative stress [95,96]. Plasma levels have been shown to decrease with increasing tumor stage in breast and cervical cancer patients [97].

Pharmacokinetic studies in cancer patients reported peak plasma levels of up to 26 mM following infusions with 1.5 g ascorbate per kilogram body weight [98] or 21 mM following 60 g ascorbate per patient [99]. Concentrations of ascorbate above 10 mM were sustained for over 4 hours following a single infusion [98]. Plasma ascorbate followed first-order elimination kinetics with a reported half-life of 1.7–2.5 hours [100] and 1.87 hours [99].

The diffusion and intracellular uptake of ascorbate in solid tumors has been predicted using an *in vitro* multicellular layer model and pharmacokinetic modeling [101]. Data showed that ascorbate diffused slowly through multicellular tissue and may follow a similar pattern to that of oxygen. According to this model, continuously elevated plasma concentrations above the physiologic maximum (>100 μM) may be required for effective delivery of ascorbate to poorly vascularized regions of the tumor [101].

TISSUE ASCORBATE LEVELS IN PRECLINICAL MODELS AND CANCER PATIENTS

Evidence suggests that accumulation of ascorbate in tumor tissue may be different from normal tissue. A pharmacokinetic study in tumor-bearing Gulo$^{-/-}$ mice (ascorbate-dependent knockout mice) has shown that in response to a single high-dose ascorbate administration, ascorbate levels in the tumor remained elevated for 48 hours, whereas liver ascorbate levels reduced rapidly along with plasma levels [102]. However, ascorbate accumulation in patient tumors following high-dose administration has not yet been reported. Instead, indirect evidence from patient tumor tissue supports an unusual pattern of ascorbate accumulation. Analysis of tissue samples from patients with colorectal cancer, endometrial cancer, breast cancer, and ccRCC showed no association between ascorbate levels measured in tumor and patient-matched normal tissue [103–106]. Only in papillary RCC tumors was ascorbate content positively correlated with normal tissue content [105].

A number of retrospective studies have measured tissue ascorbate content in cancer patients, with data showing a considerable range in ascorbate levels among patients and across cancer types. In colorectal [104,107], endometrial [103], and brain tissues [108], tumors contained significantly less ascorbate than normal tissue. In addition, in both endometrial and colorectal cancer, low tumor ascorbate levels were associated with higher tumor grade, increased necrosis, and larger tumors [103,104]. In contrast, lung cancer [95], papillary RCC, and high-grade ccRCC tumors [105] had significantly increased ascorbate levels compared to normal tissue. In patients with breast cancer, tumor ascorbate levels were reportedly similar [106] or higher [109] than normal tissue. Low ascorbate content was associated with higher-grade tumors and ductal carcinoma *in situ* necrosis in breast cancer patients [106]. In thyroid tissue, there was no difference in ascorbate content between thyroid lesions by stage or type (nodular goiters, follicular adenomas, and papillary or follicular carcinomas), and normal tissue was not measured [86].

ASSOCIATION OF HYPOXIA-INDUCIBLE FACTORS AND ASCORBATE IN PRECLINICAL MODELS AND CANCER PATIENTS

In most laboratory mice, including SCID mice that grow human xenografts, endogenous plasma and tissue levels of ascorbate are saturating as these animals synthesize ascorbate in their liver [110]. Therefore, although several *in vivo* studies have investigated ascorbate in cancer [111–113], only a handful have used relevant ascorbate-dependent animals, including guinea pigs [114] and $Gulo^{-/-}$ knockout mice [88,102,115].

Elevating circulating plasma levels of $Gulo^{-/-}$ mice via dietary intervention [88,115] or by intraperitoneal injection with high-dose ascorbate [102] elevated tumor levels of ascorbate, reduced tumor take-rate and tumor growth. Ascorbate intervention also reduced hypoxia, HIF-1α protein levels, and downstream target proteins [88,102,115]. These *in vivo* findings provide solid evidence that manipulation of tumor ascorbate levels was able to modify HIF pathway activity, but data from clinical intervention trials are not yet published.

Clinical data from (ascorbate-) untreated patients have demonstrated associations between tumor ascorbate levels and the HIF pathway. In endometrial, colorectal, thyroid, papillary RCC, and breast tumors, an inverse correlation between tumor ascorbate content and HIF pathway activation has been demonstrated [86,103–106] (Figure 4.3). HIF-1 pathway activity was reduced in tumor samples with increased ascorbate

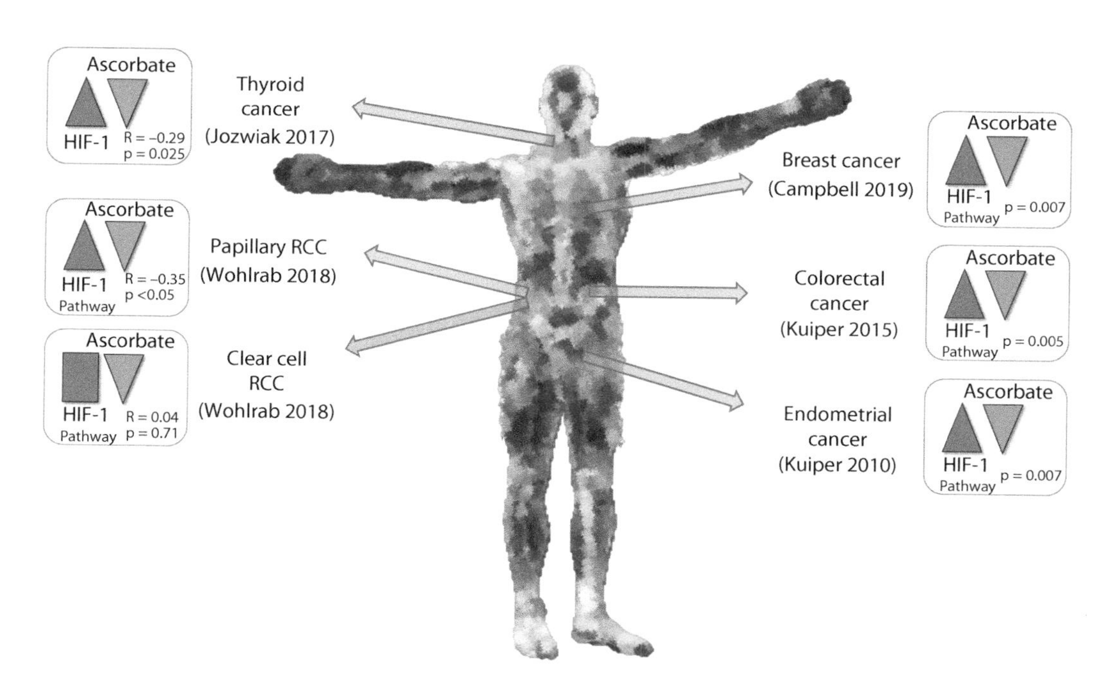

Figure 4.3. Relationship of HIF-1 and ascorbate in human cancer tissues. An inverse relationship between intracellular ascorbate levels and HIF-1/HIF-pathway is evident in thyroid, breast, colorectal, and endometrial cancers, and papillary RCC, but not in clear cell RCC, which has mutated VHL that are unable to degrade hydroxylated HIF-1. (HIF, hypoxia-inducible factor; RCC, renal cell carcinoma.) Information collated from [86,103–106].

content. HIF-2 pathway activation also tended to be associated with tumor ascorbate content in papillary RCC, whereas HIF-2α levels did not correlate with ascorbate content in thyroid cancer specimens [86,105]. Only in ccRCC, which lacks a functional VHL protein resulting in constitutive activation of HIF, was there no association between ascorbate levels and HIF pathway activity [105]. The observed inverse association between ascorbate and HIF activity in clinical samples is likely attributable to ascorbate's role as a cofactor for regulatory HIF hydroxylases. Future clinical intervention studies will determine whether increasing tumor ascorbate levels, potentially via high-dose infusion, can dampen the HIF pathway response, and whether this has clinical value.

In patients with colorectal cancer, increased levels of tumor ascorbate were associated with significantly increased disease-free survival [104]. Patients with above median tumor ascorbate levels had an additional 5–36 disease-free months in the 6 years post-cancer surgery. Ascorbate levels of matched normal bowel tissue were not associated with survival. In patients with breast cancer, disease-free and disease-specific survival was significantly improved in the top third of patients compared to the lowest third according to tumor ascorbate levels [106]. These data provide evidence that it is worth considering elevating tumor ascorbate levels in an attempt to improve poor patient survival associated with pathologic HIF pathway activity.

CONCLUSIONS

Tumor hypoxia and HIF pathway activity are associated with poor response to treatment and reduced survival in many cancer patients. Increased intracellular levels of ascorbate have been linked to reduced HIF pathway activity in cell culture, animal models, and retrospective cancer patient samples. In patients with colorectal and breast cancer, increased tumor levels of ascorbate were associated with improved survival. It still needs to be determined whether manipulation of plasma ascorbate levels, via dietary or high-dose infusion intervention, leads to increased tumor ascorbate levels, and whether this subsequently leads to a reduction in the HIF pathway response. Data from RCC patients showed that the association of ascorbate and HIF relies on a functional VHL and demonstrates that some tumor types, or some patients, may not respond to ascorbate via modulation of the HIF pathway. As ascorbate is vital to the function of the large family of 2-oxoglutarate dependent dioxygenases, modulation of the HIF pathway is only one potential way by which ascorbate affects cancer growth and progression. Carefully controlled clinical intervention trials are now needed to identify biomarkers of response. These studies should measure not only patient survival but also tumor ascorbate levels and a range of potential downstream targets, including the HIF pathway, in order to move this field from alternative to mainstream therapy.

REFERENCES

1. Levine, M. 1986 New concepts in the biology and biochemistry of ascorbic acid. *N Engl. J. Med.* 314, 892–902.
2. Hanahan, D. and Weinberg, R. A. 2011 Hallmarks of cancer: The next generation. *Cell.* 144, 646–674.
3. Vaupel, P. and Mayer, A. 2007 Hypoxia in cancer: Significance and impact on clinical outcome. *Cancer Metastasis Rev.* 26, 225–239.
4. Wilson, W. R. and Hay, M. P. 2011 Targeting hypoxia in cancer therapy. *Nat. Rev. Cancer.* 11, 393–410.
5. Höckel, M., Schlenger, K., Aral, B., Mitze, M., Schaffer, U. and Vaupel, P. 1996 Association between tumor hypoxia and malignant progression in advanced cancer of the uterine cervix. *Cancer Res.* 56, 4509–4515.
6. Greaves, M. and Maley, C. C. 2012 Clonal evolution in cancer. *Nature.* 481, 306–313.
7. Raghunand, N., Gatenby, R. A. and Gillies, R. J. 2003 Microenvironmental and cellular consequences of altered blood flow in tumours. *Br. J. Radiol.* 76, S11–22.
8. Dewhirst, M. W., Tso, C. Y., Oliver, R., Gustafson, C. S., Secomb, T. W. and Gross, J. F. 1989 Morphologic and hemodynamic comparison of tumor and healing normal tissue microvasculature. *Int. J. Radiat. Oncol. Biol. Phys.* 17, 91–99.
9. Bristow, R. G. and Hill, R. P. 2008 Hypoxia and metabolism. hypoxia, DNA repair and genetic instability. *Nat. Rev. Cancer.* 8, 180–192.
10. Gentric, G., Mieulet, V. and Mechta-Grigoriou, F. 2016 Heterogeneity in cancer metabolism: new concepts in an old field. *Antioxid. Redox. Signal.* 26, 462–485.

11. Boucher, Y., Baxter, L. T. and Jain, R. K. 1990 Interstitial pressure gradients in tissue-isolated and subcutaneous tumors: Implications for therapy. *Cancer Res.* 50, 4478–84.
12. Majmundar, A. J., Wong, W. J. and Simon, M. C. 2010 Hypoxia-inducible factors and the response to hypoxic stress. *Mol. Cell.* 40, 294–309.
13. Kaelin, W. G. and Ratcliffe, P. J. 2008 Oxygen sensing by metazoans: The central role of the HIF hydroxylase pathway. *Mol. Cell.* 30, 393–402.
14. Dachs, G. U., Patterson, A. V., Firth, J. D., Ratcliffe, P. J., Townsend, K. M., Stratford, I. J. and Harris, A. L. 1997 Targeting gene expression to hypoxic tumor cells. *Nat. Med.* 3, 515–520.
15. Mole, D. R., Blancher, C., Copley, R. R., Pollard, P. J., Gleadle, J. M., Ragoussis, J. and Ratcliffe, P. J. 2009 Genome-wide association of hypoxia-inducible factor (HIF)-1alpha and HIF-2alpha DNA binding with expression profiling of hypoxia-inducible transcripts. *J. Biol. Chem.* 284, 16767–16775.
16. Makino, Y., Cao, R., Svensson, K., Bertilsson, G., Asman, M., Tanaka, H., Cao, Y., Berkenstam, A. and Poellinger, L. 2001 Inhibitory PAS domain protein is a negative regulator of hypoxia-inducible gene expression. *Nature.* 414, 550–554.
17. Zhang, P., Yao, Q., Lu, L., Li, Y., Chen, P.-J. and Duan, C. 2014 Hypoxia-inducible factor 3 is an oxygen-dependent transcription activator and regulates a distinct transcriptional response to hypoxia. *Cell Rep.* 6, 1110–1121.
18. Dachs, G. U. and Tozer, G. M. 2000 Hypoxia modulated gene expression: Angiogenesis, metastasis and therapeutic exploitation. *Eur. J. Cancer* 36, 1649–1660.
19. Chowdhury, R., Hardy, A. and Schofield, C. J. 2008 The human oxygen sensing machinery and its manipulation. *Chem. Soc. Rev.* 37, 1308–1319.
20. Hu, C.-J., Wang, L.-Y., Chodosh, L. A., Keith, B. and Simon, M. C. 2003 Differential roles of hypoxia-inducible factor 1alpha (HIF-1alpha) and HIF-2alpha in hypoxic gene regulation. *Mol. Cell. Biol.* 23, 9361–9374.
21. Koh, M. Y. and Powis, G. 2012 Passing the Baton: The HIF switch. *Trends Biochem. Sci.* 37, 364–372.
22. Gunaratnam, L., Morley, M., Franovic, A., de Paulsen, N., Mekhail, K., Parolin, D. A. E., Nakamura, E., Lorimer, I. A. J. and Lee, S. 2003 Hypoxia inducible factor activates the transforming growth factor-alpha/epidermal growth factor receptor growth stimulatory pathway in VHL(-/-) renal cell carcinoma cells. *J. Biol. Chem.* 278, 44966–44974.
23. Smythies, J.A., Sun M., Masson N., Salama R., Simpson P.D., Murray E., Neumann V. et al. 2019 Inherent DNA-binding specificities of the HIF-1α and HIF-2α transcription factors in chromatin. *EMBO Rep.* 20 pii: e46401.
24. Zhong, H., De Marzo, A. M., Laughner, E., Lim, M., Hilton, D. A., Zagzag, D., Buechler, P., Isaacs, W. B., Semenza, G. L. and Simons, J. W. 1999 Overexpression of hypoxia-inducible factor 1alpha in common human cancers and their metastases. *Cancer Res.* 59, 5830–5835.
25. Zhou, J., Huang, S., Wang, L., Yuan, X., Dong, Q., Zhang, D. and Wang, X. 2017 Clinical and prognostic significance of HIF-1α overexpression in oral squamous cell carcinoma: A meta-analysis. *World J. Surg. Oncol.* 15, 104.
26. Yang, S.-L., Ren, Q.-G., Wen, L. and Hu, J.-L. 2016 Clinicopathological and prognostic significance of hypoxia-inducible factor-1 alpha in lung cancer: A systematic review with meta-analysis. *J. Huazhong Univ. Sci. Technolog. Med. Sci.* 36, 321–327.
27. Ouyang, Y., Li, H., Bu, J., Li, X., Chen, Z. and Xiao, T. 2016 Hypoxia-inducible factor-1 expression predicts osteosarcoma patients' survival: A meta-analysis. *Int. J. Biol. Markers.* 31, e229–234.
28. Swartz, J. E., Pothen, A. J., Stegeman, I., Willems, S. M. and Grolman, W. 2015 Clinical implications of hypoxia biomarker expression in head and neck squamous cell carcinoma: A systematic review. *Cancer Med.* 4, 1101–1116.
29. Ye, L.-Y., Zhang, Q., Bai, X.-L., Pankaj, P., Hu, Q.-D. and Liang, T.-B. 2014 Hypoxia-inducible factor 1α expression and its clinical significance in pancreatic cancer: A meta-analysis. *Pancreatology.* 14, 391–397.
30. Huang, M., Chen, Q., Xiao, J., Yao, T., Bian, L., Liu, C. and Lin, Z. 2014 Overexpression of hypoxia-inducible factor-1α is a predictor of poor prognosis in cervical cancer: A clinicopathologic study and a meta-analysis. *Int. J. Gynecol. Cancer.* 24, 1054–1064.
31. Jin, Y., Wang, H., Liang, X., Ma, J. and Wang, Y. 2014 Pathological and prognostic significance of hypoxia-inducible factor 1α expression in epithelial ovarian cancer: A meta-analysis. *Tumour Biol.* 35, 8149–8159.
32. Cao, S., Yang, S., Wu, C., Wang, Y., Jiang, J. and Lu, Z. 2014 Protein expression of hypoxia-inducible

factor-1 alpha and hepatocellular carcinoma: A systematic review with meta-analysis. *Clin. Res. Hepatol. Gastroenterol.* 38, 598–603.

33. Jing, S. W., Wang, J. and Xu, Q. 2017 Expression of hypoxia inducible factor 1 alpha and its clinical significance in esophageal carcinoma: A meta-analysis. *Tumour Biology.* 39, 1010428317717983.
34. Wang, W., He, Y.-F., Sun, Q.-K., Wang, Y., Han, X.-H., Peng, D.-F., Yao, Y.-W., Ji, C.-S. and Hu, B. 2014 Hypoxia-inducible factor 1α in breast cancer prognosis. *Clin. Chim. Acta.* 428, 32–37.
35. Chen, Z., He X., Xia, W., Huang, Q., Zhang, Z., Ye, J., Ni, C. et al. 2013 Prognostic value and clinicopathological differences of hifs in colorectal cancer: Evidence from meta-analysis. *PLOS ONE.* 8: e80337.
36. Horrée, N., van Diest, P. J., van der Groep, P., Sie-Go, D. M. D. S. and Heintz, A. P. M. 2007 Hypoxia and angiogenesis in endometrioid endometrial carcinogenesis. *Cell. Oncol.* 29, 219–227.
37. Ozbudak, I. H., Karaveli, S., Simsek, T., Erdogan, G. and Pestereli, E. 2008 Neoangiogenesis and expression of hypoxia-inducible factor 1alpha, vascular endothelial growth factor, and glucose transporter-1 in endometrioid type endometrium adenocarcinomas. *Gynecol. Oncol.* 108, 603–608.
38. Griffiths, E. A., Pritchard, S. A., Valentine, H. R., Whitchelo, N., Bishop, P. W., Ebert, M. P., Price, P. M., Welch, I. M. and West, C. M. L. 2007 Hypoxia-inducible factor-1alpha expression in the gastric carcinogenesis sequence and its prognostic role in gastric and gastro-oesophageal adenocarcinomas. *Br. J. Cancer.* 96, 95–103.
39. Peng, G. and Liu, Y. 2015 Hypoxia-inducible factors in cancer stem cells and inflammation. *Trends Pharmacol. Sci.* 36, 374–383.
40. Sermeus, A., Cosse, J.-P., Crespin, M., Mainfroid, V., de Longueville, F., Ninane, N., Raes, M., Remacle, J. and Michiels, C. 2008 Hypoxia induces protection against etoposide-induced apoptosis: Molecular profiling of changes in gene expression and transcription factor activity. *Mol. Cancer* 7, 27.
41. Liu, L., Ning, X., Sun, L., Zhang, H., Shi, Y., Guo, C., Han, S. et al. 2008 Hypoxia-inducible factor-1 alpha contributes to hypoxia-induced chemoresistance in gastric cancer. *Cancer Sci.* 99, 121–128.
42. Rohwer, N., Dame, C., Haugstetter, A., Wiedenmann, B., Detjen, K., Schmitt, C. A. and Cramer, T. 2010 Hypoxia-inducible factor 1alpha determines gastric cancer chemosensitivity via modulation of P53 and NF-KappaB. *PLOS ONE.* 5, e12038.
43. Sullivan, R. and Graham, C. H. 2009 Hypoxia prevents etoposide-induced DNA damage in cancer cells through a mechanism involving hypoxia-inducible factor 1. *Mol. Cancer Ther.* 8, 1702–1713.
44. Sasabe, E., Zhou, X., Li, D., Oku, N., Yamamoto, T. and Osaki, T. 2007 The involvement of hypoxia-inducible factor-1alpha in the susceptibility to gamma-rays and chemotherapeutic drugs of oral squamous cell carcinoma cells. *Int. J. Cancer.* 120, 268–277.
45. Lendahl, U., Lee, K. L., Yang, H. and Poellinger, L. 2009 Generating specificity and diversity in the transcriptional response to hypoxia. *Nature Reviews. Genetics* 10, no. 12, 821–832.
46. Thomlinson, R. H. and Gray, L. H. 1955 The histological structure of some human lung cancers and the possible implications for radiotherapy. *Br. J. Cancer.* 9, 539–549.
47. Williams, K. J., Telfer, B. A., Xenaki, D., Sheridan, M. R., Desbaillets, I., Peters, H. J. W., Honess, D. et al. 2005 Enhanced response to radiotherapy in tumours deficient in the function of hypoxia-inducible factor-1. *Radiother. Oncol.* 75, 89–98.
48. Huang, Y., Yu, J., Yan, C., Hou, J., Pu, J., Zhang, G., Fu, Z. and Wang, X. 2012 Effect of small interfering RNA targeting hypoxia-inducible factor-1α on radiosensitivity of PC3 cell line. *Urology.* 79, 744.e17–24.
49. Liu, J., Zhang, J., Wang, X., Li, Y., Chen, Y., Li, K., Zhang, J., Yao, L. and Guo, G. 2010 HIF-1 and NDRG2 contribute to hypoxia-induced radioresistance of cervical cancer Hela cells. *Exp. Cell Res.* 316, 1985–1993.
50. Schilling, D., Bayer, C., Emmerich, K., Molls, M., Vaupel, P., Huber, R. M. and Multhoff, G. 2012 Basal HIF-1α expression levels are not predictive for radiosensitivity of human cancer cell lines. *Strahlenther. Onkol.* 188, 353–358.
51. Xia, Y., Jiang, L. and Zhong, T. 2018 The role of HIF-1α in chemo-/radioresistant tumors. *Onco. Targets. Ther.* 11, 3003–3011.
52. Stover, E. H., Konstantinopoulos, P. A., Matulonis, U. A. and Swisher, E. M. 2016 Biomarkers of response and resistance to DNA repair targeted therapies. *Clin. Cancer Res.* 22, 5651–5660.
53. Rezaeian, A.-H., Wang, Y.-H. and Lin, H.-K. 2017 DNA damage players are linked to HIF-1α/hypoxia signaling. *Cell Cycle.* 16, 725–726.

54. Jiang, B. H., Zheng, J. Z., Leung, S. W., Roe, R. and Semenza, G. L. 1997 Transactivation and inhibitory domains of hypoxia-inducible factor 1alpha. Modulation of transcriptional activity by oxygen tension. *J. Biol. Chem.* 272, 19253–19260.
55. Li, H., Ko, H. P. and Whitlock, J. P. 1996 Induction of phosphoglycerate kinase 1 gene expression by hypoxia ROLESInduction of phosphoglycerate kinase 1 gene expression by hypoxia. *Roles of Arnt and HIF1alpha. Journal of Biological Chemistry.* 271, 21262–21267.
56. Hu, C.-J., Sataur, A., Wang, L., Chen, H., Simon, M. C. and Tansey, W. 2007 The N-terminal transactivation domain confers target gene specificity of hypoxia-inducible factors HIF-1α and HIF-2α. *Mol. Biol. Cell.* 18, 4528–4542.
57. Huang, L. E., Gu, J., Schau, M. and Bunn, H. F. 1998 Regulation of hypoxia-inducible factor 1alpha is mediated by an O2-dependent degradation domain via the ubiquitin-proteasome pathway. *Proc. Natl. Acad. Sci. USA* 95, 7987–7992.
58. Pugh, C. W. and Ratcliffe, P. J. 2017 New Horizons in hypoxia signaling pathways. *Exp. Cell Res.* 356, 116–121.
59. Vissers, M. C. M., Kuiper, C. and Dachs, G. U. 2014 Regulation of the 2-oxoglutarate-dependent dioxygenases and implications for cancer. *Biochem. Soc. Trans.* 42, 945–951.
60. Lando, D., Peet, D. J., Whelan, D. A., Gorman, J. J. and Whitelaw, M. L. 2002 Asparagine hydroxylation of the HIF transactivation domain a hypoxic switch. *Science* 295, 858–861.
61. Berra, E., Benizri, E., Ginouvès, A., Volmat, V., Roux, D. and Pouysségur, J. 2003 HIF prolyl-hydroxylase 2 is the key oxygen sensor setting low steady-state levels of HIF-1alpha in normoxia. *EMBO J.* 22, 4082–4090.
62. Maxwell, P. H., Wiesener, M. S., Chang, G. W., Clifford, S. C., Vaux, E. C., Cockman, M. E., Wykoff, C. C., Pugh, C. W., Maher, E. R. and Ratcliffe, P. J. 1999 The tumour suppressor protein VHL targets hypoxia-inducible factors for oxygen-dependent proteolysis. *Nature.* 399, 271–275.
63. Appelhoff, R. J., Tian, Y.-M., Raval, R. R., Turley, H., Harris, A. L., Pugh, C. W., Ratcliffe, P. J. and Gleadle, J. M. 2004 Differential function of the prolyl hydroxylases PHD1, PHD2, and PHD3 in the regulation of hypoxia-inducible factor. *J. Biol. Chem.* 279, 38458–38465.
64. Koivunen, P., Hirsilä, M., Günzler, V., Kivirikko, K. I. and Myllyharju, J. 2004 Catalytic properties of the asparaginyl hydroxylase (FIH) in the oxygen sensing pathway are distinct from those of its prolyl 4-hydroxylases. *J. Biol. Chem.* 279, 9899–9904.
65. Hirsilä, M., Koivunen, P., Günzler, V., Kivirikko, K. I. and Myllyharju, J. 2003 Characterization of the human prolyl 4-hydroxylases that modify the hypoxia-inducible factor. *J. Biol. Chem.* 278, 30772–30780.
66. Tian, Y.-M., Yeoh, K. K., Lee, M. K., Eriksson, T., Kessler, B. M., Kramer, H. B., Edelmann, M. J. et al. 2011 Differential sensitivity of hypoxia inducible factor hydroxylation sites to hypoxia and hydroxylase inhibitors. *J. Biol. Chem.* 286, 13041–13051.
67. Khan, M. N., Bhattacharyya, T., Andrikopoulos, P., Esteban, M. A., Barod, R., Connor, T., Ashcroft, M., Maxwell, P. H. and Kiriakidis, S. 2011 Factor inhibiting HIF (FIH-1) promotes renal cancer cell survival by protecting cells from HIF-1α-mediated apoptosis. *Br. J. Cancer.* 104, 1151–1159.
68. LaGory, E. L. and Giaccia, A. J. 2016 The ever-expanding role of HIF in tumour and stromal biology. *Nat. Cell Biol.* 18, 356–365.
69. Hon, W.-C., Wilson, M. I., Harlos, K., Claridge, T. D. W., Schofield, C. J., Pugh, C. W., Maxwell, P. H., Ratcliffe, P. J., Stuart, D. I. and Jones, E. Y. 2002 Structural basis for the recognition of hydroxyproline in HIF-1α by PVHL. *Nature.* 417, 975–978.
70. Nguyen, H. C., Yang, H., Fribourgh, J. L., Wolfe, L. S. and Xiong, Y. 2015 Insights into Cullin-RING E3 ubiquitin ligase recruitment: Structure of the VHL–EloBC–Cul2 complex. *Structure* 23, 441–449.
71. Gossage, L., Eisen, T. and Maher, E. R. 2015 VHL, the story of a tumour suppressor gene. *Nature Reviews. Cancer.* 15, 55–64.
72. Koh, M. Y., Nguyen, V., Lemos, R., Darnay, B. G., Kiriakova, G., Abdelmelek, M., Ho, T. H. et al. 2015 Hypoxia-induced SUMOylation of E3 ligase HAF determines specific activation of HIF2 in clear-cell renal cell carcinoma. *Cancer Res.* 75, 316–329.
73. Ozer, A. and Bruick, R. K. 2007 Non-heme dioxygenases: cellular sensors and regulators jelly rolled into one? *Nat. Chem. Biol.* 3, 144–153.
74. Zhou, J., Gunsior, M., Bachmann, B. O., Townsend, C. A. and Solomon, E. I. 1998 Substrate binding to the α-ketoglutarate-dependent non-heme iron enzyme clavaminate synthase 2: Coupling mechanism of oxidative

decarboxylation and hydroxylation. *J. Am. Chem. Soc.* 120, 13539–13540.
75. Kuiper, C. and Vissers, M. C. M. 2014 Ascorbate as a co-factor for Fe- and 2-oxoglutarate dependent dioxygenases: Physiological activity in tumor growth and progression. *Front. Oncol* 4, 359.
76. Hoffart, L. M., Barr, E. W., Guyer, R. B., Bollinger, J. M. and Krebs, C. 2006 Direct spectroscopic detection of a C-H-cleaving high-spin Fe(IV) complex in a prolyl-4-hydroxylase. *Proc. Natl. Acad. Sci. USA* 103, 14738–14743.
77. Flashman, E., Davies, S. L., Yeoh, K. K. and Schofield, C. J. 2010 Investigating the dependence of the hypoxia-inducible factor hydroxylases (factor inhibiting HIF and prolyl hydroxylase domain 2) on ascorbate and other reducing agents. *Biochem. J.* 427, 135–142.
78. Clifton, I. J., McDonough, M. A., Ehrismann, D., Kershaw, N. J., Granatino, N. and Schofield, C. J. 2006 Structural studies on 2-oxoglutarate oxygenases and related double-stranded beta-helix fold proteins. *J. Inorg. Biochem.* 100, 644–669.
79. Osipyants, A. I., Poloznikov, A. A., Smirnova, N. A., Hushpulian, D. M., Khristichenko, A. Y., Chubar, T. A., Zakhariants, A. A. et al. 2018 L-ascorbic acid: A true substrate for HIF prolyl hydroxylase? *Biochimie.* 147, 46–54.
80. Wohlrab, C., Kuiper, C., Vissers, M. C., Phillips, E., Robinson, B. A. and Dachs, G. U. 2019 Ascorbate modulates the hypoxic pathway by increasing intracellular activity of the HIF hydroxylases in renal cell carcinoma cells. *Hypoxia*, in press.
81. Kuiper, C., Dachs, G. U., Currie, M. J. and Vissers, M. C. M. 2014 Intracellular ascorbate enhances hypoxia-inducible factor (HIF)-hydroxylase activity and preferentially suppresses the HIF-1 transcriptional response. *Free Radic. Biol. Med.* 69, 308–317.
82. Knowles, H. J., Raval, R. R., Harris, A. L. and Ratcliffe, P. J. 2003 Effect of ascorbate on the activity of hypoxia-inducible factor in cancer cells. *Cancer Res.* 63, 1764–1768.
83. Lu, H., Dalgard, C. L., Mohyeldin, A., McFate, T., Tait, A. S. and Verma, A. 2005 Reversible inactivation of HIF-1 prolyl hydroxylases allows cell metabolism to control Basal HIF-1. *J. Biol. Chem.* 280, 41928–41939.
84. Vissers, M. C. M., Gunningham, S. P., Morrison, M. J., Dachs, G. U. and Currie, M. J. 2007 Modulation of hypoxia-inducible factor-1 alpha in cultured primary cells by intracellular ascorbate. *Free Radic. Biol. Med.* 42, 765–772.
85. Qiao, H. and May, J. M. 2009 Macrophage differentiation increases expression of the ascorbate transporter (SVCT2). *Free Radic. Biol. Med.* 46, 1221–1232.
86. Jóźwiak, P., Ciesielski, P., Zaczek, A., Lipińska, A., Pomorski, L., Wieczorek, M., Bryś, M., Forma, E. and Krześlak, A. 2017 Expression of hypoxia inducible factor 1α and 2α and its association with vitamin C level in thyroid lesions. *J. Biomed. Sci.* 24, 83.
87. Fischer, A. P. and Miles, S. L. 2017 Ascorbic acid, but not dehydroascorbic acid increases intracellular vitamin C content to decrease hypoxia inducible factor -1 alpha activity and reduce malignant potential in human melanoma. *Biomed. Pharmacother.* 86, 502–513.
88. Campbell, E. J., Vissers, M. C. and Dachs, G. U. 2016 Ascorbate availability affects tumor implantation-take rate and increases tumor rejection in Gulo-/-mice. *Hypoxia.* 4, 41–52.
89. Scheers, N. and Sandberg, A. S. 2014 Iron transport through ferroportin is induced by intracellular ascorbate and involves IRP2 and HIF2α. *Nutrients.* 6, 249–60.
90. Wohlrab, C., Phillips, E. and Dachs, G. U. 2017 Vitamin C transporters in cancer: Current understanding and gaps in knowledge. *Front. Oncol.* 7, 74–0.
91. Padayatty, S. J., Sun, H., Wang, Y., Riordan, H. D., Hewitt, S. M., Katz, A., Wesley, R. A. and Levine, M. 2004 Vitamin C pharmacokinetics: Implications for oral and intravenous use. *Ann. Intern. Med.* 140, 533–537.
92. Ros, M. M., Bueno-de-Mesquita, H. B., Kampman, E., Aben, K. K. H., Büchner, F. L., Jansen, E. H. J. M., van Gils, C. H. et al. 2012 Plasma carotenoids and vitamin C concentrations and risk of urothelial cell carcinoma in the European prospective investigation into cancer and nutrition. *Am. J. Clin. Nutr.* 96, 902–910.
93. Nagamma, T., Baxi, J. and Singh, P. P. 2014 Status of oxidative stress and antioxidant levels in smokers with breast cancer from Western Nepal. *Asian Pac. J. Cancer Prev.* 15, 9467–9470.
94. Gackowski, D., Kowalewski, J., Siomek, A. and Olinski, R. 2005 Oxidative DNA damage and antioxidant vitamin level: Comparison among

lung cancer patients, healthy smokers and nonsmokers. *Int. J. Cancer*. 114, 153–156.

95. Anthony, H. M. and Schorah, C. J. 1982 Severe hypovitaminosis C in lung-cancer patients: The utilization of vitamin C in surgical repair and lymphocyte-related host resistance. *Br. J. Cancer* 46, 354–367.
96. Mayland, C. R., Bennett, M. I. and Allan, K. 2005 Vitamin C deficiency in cancer patients. *Palliat. Med*. 19, 17–20.
97. Ramaswamy, G. and Krishnamoorthy, L. 1996 Serum carotene, vitamin A, and vitamin C levels in breast cancer and cancer of the uterine cervix. *Nutr. Cancer*. 25, 173–177.
98. Hoffer, L. J., Levine, M., Assouline, S., Melnychuk, D., Padayatty, S. J., Rosadiuk, K., Rousseau, C., Robitaille, L. and Miller, W. H. 2008 Phase I clinical trial of i.v. Ascorbic acid in advanced malignancy. *Ann. Oncol*. 19, 1969–1974.
99. Nielsen, T. K., Højgaard, M., Andersen, J. T., Poulsen, H. E., Lykkesfeldt, J. and Mikines, K. J. 2015 Elimination of ascorbic acid after high-dose infusion in prostate cancer patients: A pharmacokinetic evaluation. *Basic Clin. Pharmacol. Toxicol*. 116, 343–8.
100. Stephenson, C. M., Levin, R. D., Spector, T. and Lis, C. G. 2013 Phase I clinical trial to evaluate the safety, tolerability, and pharmacokinetics of high-dose intravenous ascorbic acid in patients with advanced cancer. *Cancer Chemother. Pharmacol*. 72, 139–46.
101. Kuiper, C., Vissers, M. C. M. and Hicks, K. O. 2014 Pharmacokinetic modeling of ascorbate diffusion through normal and tumor tissue. *Free Radic. Biol. Med*. 77, 340–352.
102. Campbell, E. J., Vissers, M. C. M., Wohlrab, C., Hicks, K. O., Strother, R. M., Bozonet, S. M., Robinson, B. A. and Dachs, G. U. 2016 Pharmacokinetic and anti-cancer properties of high dose ascorbate in solid tumours of ascorbate-dependent mice. *Free Radic. Biol. Med*. 99, 451–462.
103. Kuiper, C., Molenaar, I. G. M., Dachs, G. U., Currie, M. J., Sykes, P. H. and Vissers, M. C. M. 2010 Low Ascorbate levels are associated with increased hypoxia-inducible factor-1 activity and an aggressive tumor phenotype in endometrial cancer. *Cancer Res*. 70, 5749–5758.
104. Kuiper, C., Dachs, G. U., Munn, D., Currie, M. J., Robinson, B. A., Pearson, J. F. and Vissers, M. C. M. 2014 Increased tumor ascorbate is associated with extended disease-free survival and decreased hypoxia-inducible factor-1 activation in human colorectal cancer. *Front. Oncol*. 4, 10.
105. Wohlrab, C., Vissers, M. C. M., Phillips, E., Morrin, H., Robinson, B. A. and Dachs, G. U. 2018 The association between ascorbate and the hypoxia-inducible factors in human renal cell carcinoma requires a functional Von Hippel-Lindau protein. *Front. Oncol*. 8, 574.
106. Campbell, E.J., G.U. Dachs, H. Morrin, V. Davey, B.A. Robinson, and M.C.M. Vissers. 2019 Activation of the hypoxia pathway in breast cancer tissue and patient survival are inversely associated with tumor ascorbate levels. *BMC Cancer*. 19(1), 307.
107. Skrzydlewska, E., Stankiewicz, A., Sulkowska, M., Sulkowski, S. and Kasacka, I. 2001 Antioxidant status and lipid peroxidation in colorectal cancer. *J Toxicol Environ Health A*. 64, 213–222.
108. Dudek, H., Farbiszewski, R., Rydzewska, M., Michno, T. and Kozłowski, A. 2004 Evaluation of antioxidant enzymes activity and concentration of non-enzymatic antioxidants in human brain tumours. *Wiad. Lek. (Warsaw, Poland: 1960)*. 57, 16–19.
109. Langemann, H., Torhorst, J., Kabiersch, A., Krenger, W. and Honegger, C. G. 1989 Quantitative determination of water- and lipid-soluble antioxidants in neoplastic and non-neoplastic human breast tissue. *Int. J. Cancer*. 43, 1169–1173.
110. Campbell, E. J. and Dachs, G. U. 2014 Current limitations of murine models in oncology for ascorbate research. *Front. Oncol*. 4, 282.
111. Gao, P., Zhang, H., Dinavahi, R., Li, F., Xiang, Y., Raman, V., Bhujwalla, Z. M. et al. 2007 HIF-dependent antitumorigenic effect of antioxidants in vivo. *Cancer Cell*. 12, 230–238.
112. Chen, Q., Espey, M. G., Sun, A. Y., Lee, J.-H., Krishna, M. C., Shacter, E., Choyke, P. L. et al. 2007 Ascorbate in pharmacologic concentrations selectively generates ascorbate radical and hydrogen peroxide in extracellular fluid in vivo. *Proc. Natl. Acad. Sci. USA* 104, 8749–8754.
113. Chen, Q., Espey, M. G., Sun, A. Y., Pooput, C., Kirk, K. L., Krishna, M. C., Khosh, D. B., Drisko, J. and Levine, M. 2008 Pharmacologic doses of ascorbate act as a prooxidant and decrease growth of aggressive tumor xenografts in mice. *Proc. Natl. Acad. Sci. USA* 105, 11105–11109.

114. Buehler, P. W., D'Agnillo, F., Hoffman, V. and Alayash, A. I. 2007 Effects of endogenous ascorbate on oxidation, oxygenation, and toxicokinetics of cell-free modified hemoglobin after exchange transfusion in rat and guinea pig. *J. Pharmacol. Exp. Ther.* 323, 49–60.

115. Campbell, E. J., Vissers, M. C. M., Bozonet, S., Dyer, A., Robinson, B. A. and Dachs, G. U. 2015 Restoring physiological levels of ascorbate slows tumor growth and moderates HIF-1 pathway activity in Gulo(-/-) mice. *Cancer Med.* 4, 303–314.

CHAPTER FIVE

Pharmacologic Ascorbate as a Radiosensitizer

Joseph J. Cullen and Matthew S. Alexander

CONTENTS

INTRODUCTION

Pancreatic cancer is the fourth leading cause of cancer death in the United States [1]. While there is a significant effort devoted to identifying novel systemic agents, there is a tremendous clinical need to improve local therapies. Indeed, a rapid autopsy series to analyze causes of death in pancreatic ductal adenocarcinoma (PDAC) patients demonstrated that ~30% of patients died due to the consequences of local disease [2]. Surgery can be used as a local therapy, but the majority of patients have disease that is too advanced to allow for surgical resection. The alternative local therapy is utilizing radiation either alone or in combination with chemotherapy. Despite increasing the intensity of treatment by altering radiation fractionation or with different concurrent chemotherapies, most patients will experience local failure and succumb to their disease. Thus, there are a significant number of patients who would benefit from improvements in the efficacy of radio- and chemotherapy for pancreatic cancer.

Clinical data show that when ascorbate is given orally, fasting plasma concentrations are tightly controlled at less than 100 μM [3,4]. In contrast, when ascorbate is administered intravenously, plasma concentrations as high as 1–30 mM are safely achieved as seen in the phase I trial conducted at the University of Iowa (NCT 01049880, Cullen PI) demonstrating that pharmacologic ascorbate (high-dose, intravenous, vitamin C, P-AscH$^-$) is safe, is well tolerated, and has potential efficacy [3]. Thus, it is clear that intravenous administration of ascorbate can yield very high plasma levels, while oral treatment does not. Ascorbate-mediated cell death has been shown to be due to H_2O_2 generation, via ascorbate radical formation, with ascorbate as the electron donor (Figure 5.1) [5]. When ascorbate is infused intravenously, the resulting pharmacologic concentration distributes rapidly into the extracellular water space, generating ascorbate radical and H_2O_2 [6]. In contrast, the same pharmacologic ascorbate concentrations in whole blood generate little detectable ascorbate radical and no detectable H_2O_2 [7]. This can be accounted for by efficient and redundant H_2O_2 catabolic pathways in whole blood relative to those in media or extracellular fluid [8]. Thus, ascorbate administered intravenously in pharmacologic concentrations may serve as a prodrug for H_2O_2 delivery to the extracellular milieu, but without significant H_2O_2 accumulation in blood. Our *in vitro*, *in vivo*, and human studies combined with those of others provide a solid foundation for using P-AscH$^-$ as a radiosensitizer in PDAC therapy [5–7,9,10]. Our recent, first in-human phase

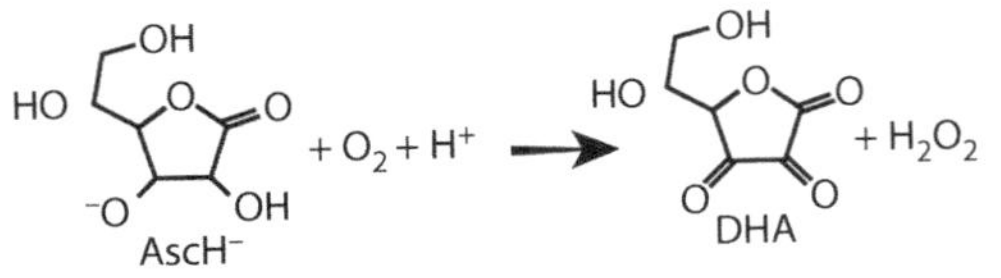

$AscH^- + O_2 + H^+ \rightarrow DHA + H_2O_2$

Figure 5.1. Ascorbate oxidation generates H_2O_2. At pharmacologic concentrations, ascorbate can act as a pro-oxidant by donating an electron to oxygen, ultimately generating H_2O_2 and dehydroascorbic acid (DHA).

I study demonstrated that P-AscH$^-$ combined with chemoradiation is safe and well tolerated and may lead to overall clinical benefit in patients with pancreatic ductal adenocarcinoma.

The generation of reactive oxygen species (ROS) is a major mechanism responsible for the therapeutic effect of ionizing radiation and of nearly 50% of chemotherapeutic drugs [11]. Currently, these therapeutic strategies are being employed to kill cancer cells without the benefit of a rational design that exploits the intrinsic differences in the cellular redox status between normal and cancer cells. Pancreatic cancer cells are under a higher oxidative stress than normal cells, and an additional increase in pro-oxidant levels can trigger cell death as demonstrated in Figure 5.2 [12,13]. The spectrum of cellular responses to oxidative stress is variable. Exposure of dividing cells in culture to low concentrations of oxidants (e.g., a pulse of 3–15 μM H_2O_2) actually stimulates cell growth and division [14]. At higher concentrations of H_2O_2 (>250 μM), cells will undergo growth arrest and may become necrotic [15]. Our *in vitro* studies demonstrate that we achieve H_2O_2 concentrations of 250 μM with 10 mM ascorbate [5]. *In vivo*, we can achieve concentrations of ascorbate up to 50 mM in the serum [3]. Our working model is that this high level of ascorbate will lead to a high flux of H_2O_2, resulting in tumor cell killing.

Another way to depict this is seen in Figure 5.2. The redox environment of a cell changes throughout its life cycle [16]. During proliferation, the Eh_c of glutathione, the principal component of the intracellular redox buffer, has the most negative value (i.e., is in the most reduced state). In this low/reducing state, the switches for proliferation are fully on. Our hypothesis is that the H_2O_2 produced by P-AscH$^-$ pushes the cellular redox buffer in tumor cells to a much more oxidized state (i.e., more positive). This results in growth arrest and tumor cell killing. When Eh_c is more positive (a), the adapted signaling differentiation switches can be turned on in normal cells for physiologic signaling. The more positive Eh_c becomes (b), the more adaptive signaling is turned on in tumor cells until they reach a maximum where nearly all cells are proliferating (c). But in (d), the redox buffer has become more oxidized; if Eh_c becomes too positive, then death signals are activated, and necrosis is initiated. This mechanism provides for the orderly removal of cells that have lost their ability to control their redox environment and, therefore, are not functioning normally. Very high values of Eh_c resulting from severe oxidative stress leave only necrosis as a path to cell death. Our hypothesis is that pancreatic tumor cell death will occur with high-dose ascorbate because the high flux of H_2O_2 will push the redox buffer to a very oxidized state that is not compatible with cell survival.

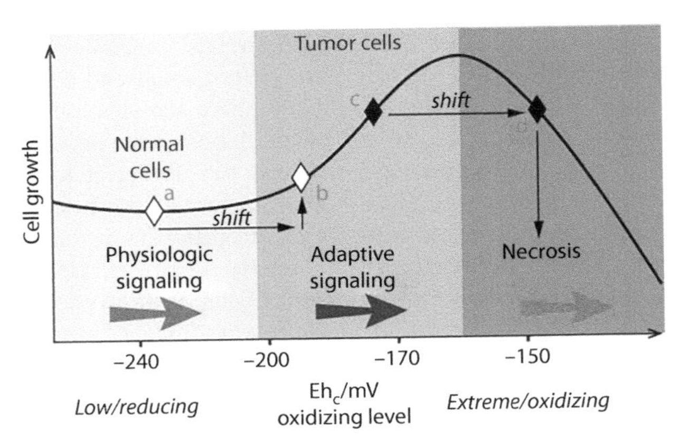

Figure 5.2. The concept of combining P-AscH$^-$ with ionizing radiation in terms of the redox homeostasis of the cell. Normal cells that are at a lower oxidizing (reducing) state will be resistant to P-AscH$^-$-induced oxidative stress compared to cancer cells that are at a higher oxidation state.

Our previous studies have also demonstrated that pancreatic cancer cells with mutant K-ras have increased levels of ROS [13] and may be in more of an oxidized state. To determine if K-ras increases ROS, we used an immortalized cell line derived from normal pancreatic ductal epithelial with near-normal genotype and phenotype of pancreatic duct epithelial cells and the isogenic cell line that expresses K-ras and forms tumors. Both dihydroethidium fluorescence and dichloro-dihydro-fluorescein fluorescence were increased when K-ras was overexpressed. The K-ras-induced increases in ROS were not due to changes in the antioxidant capacity that detoxifies superoxide, since manganese superoxide dismutase and copper/zinc dismutase levels were unchanged with K-ras overexpression. To further support this concept, we measured steady-state levels of superoxide in resected pancreatic cancer specimens versus normal pancreas (Figure 5.3). Steady-state levels of superoxide as measured by dihydroethidium were increased in pancreatic cancer specimens compared to normal pancreas. Thus, therapeutic approaches that use pro-oxidants like P-AscH$^-$ to push tumor cells into oxidative stress overload but stimulate adaptive responses in normal cells can be developed to selectively enhance radiation therapy efficacy in killing cancer cells.

PHARMACOLOGIC ASCORBATE (P-ASCH$^-$) AS A RADIOSENSITIZER

Our studies have demonstrated that P-AscH$^-$ is the prototypical antioxidant/pro-oxidant that can elicit a protective response in normal cells but add injury to tumor cells after radiation therapy [17]. P-AscH$^-$ enhanced the cytotoxic effects of radiation as seen by decreased cell viability and clonogenic survival in all pancreatic cancer cell lines examined, but not in nontumorigenic pancreatic ductal epithelial cells. P-AscH$^-$ radiosensitization was associated with an increase in oxidative stress-induced DNA damage, which was reversed by catalase. In mice with established heterotopic and orthotopic pancreatic cancer xenografts, P-AscH$^-$ combined with radiation decreased tumor growth and increased survival. Most importantly, P-AscH$^-$ radiosensitized pancreatic cancer tumors *in vivo*, reversed radiation-induced damage to the gastrointestinal tract, and did not increase systemic changes in parameters indicative of oxidative stress [17]. These results clearly demonstrate the potential clinical utility of P-AscH$^-$ as a radiosensitizer in the treatment of pancreatic cancer.

These very encouraging translational studies led us to perform the first in human trial "Gemcitabine, Ascorbate, Radiation Therapy for PDAC, Phase I" (NCT01852890). Work done by ECOG Trial 4201 showed that overall survival (OS) was 11.1 months for gemcitabine combined with chemoradiation versus 9.2 months for gemcitabine alone [18]. Approved by the U.S. Food and Drug Administration and the University of Iowa IRB-01, the trial was listed on clinicaltrials.gov [NCT01852890]. The P-AscH$^-$ dose was escalated *via* Storer's Phase I Two-Stage Design BD [19]. There were three planned cohorts: 50 g, 75 g, and 100 g. Patients were required to have histologically or cytologically diagnosed pancreatic cancer and

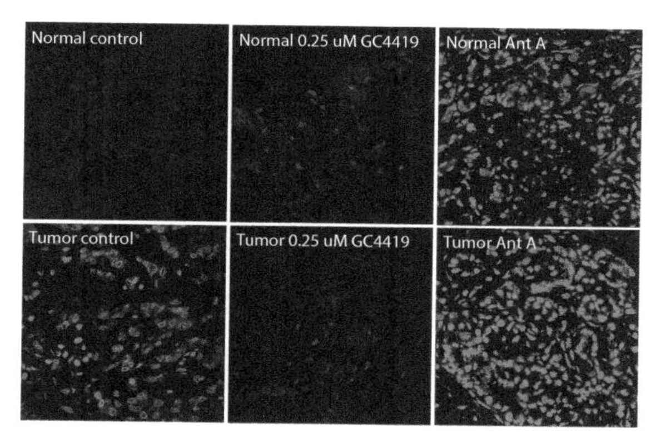

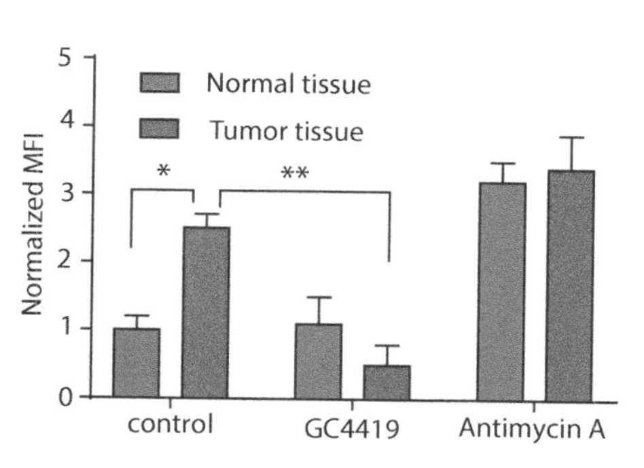

Figure 5.3. Experimental evidence of differences in oxidation state of pancreatic cancer tissues compared to adjacent nonmalignant tissues. Microscopy was used to score mean fluorescence intensity (MFI) of dihydroethidium (DHE) oxidation of tissue sections. Quantitative results are shown on right. Antimycin was used as a positive control, and GC4419 (superoxide dismutase mimetic) was used to show specificity of DHE for detecting reactive oxygen species (superoxide). ($N = 3$; $^*p < 0.05$ normal vs. tumor; $^{**}p < 0.05$ tumor tissue before and after addition of superoxide dismutase mimetic.)

indication for chemoradiotherapy. Gemcitabine was administered with an intravenous infusion at a dose of 600 mg/m^2 over 30 minutes, once weekly for 6 weeks [18]. Radiation was planned per standard of care guidelines (NCCN and RTOG). Patients were treated with either 50.4 Gy in 28 fractions or 50 Gy in 25 fractions as determined most appropriate by the treating radiation oncologist. Fourteen patients completed all protocol therapy. No subjects were lost to follow-up. A comparator cohort of 21 patients at our institution with pancreatic cancer who received gemcitabine and radiation were also evaluated. This first in human phase I trial demonstrated that P-AscH$^-$ in combination with gemcitabine and radiation for locally advanced pancreatic cancer is safe and well tolerated with suggestions of efficacy (Figure 5.4) [20]. The median OS of patients treated with P-AscH$^-$, gemcitabine, and radiation was 21.7 months (Figure 5.4a) compared to a median OS of 12.7 months in the comparator arm of the study ($p = .05$). This rate is also increased compared to historical median OS of 11.1 months published by Loehrer et al. [18]. Median progression-free survival (PFS) was 13.7 months in the P-AscH$^-$-treated patients and 4.6 months in the comparator patients treated with gemcitabine and radiation ($p = .01$) (Figure 5.4b).

As seen in the phase I trial, pharmacologic concentrations of ascorbate produce H_2O_2 via the formation of ascorbate radical (Asc$^{\bullet-}$) [3] (Figure 5.5).

PHARMACOLOGIC ASCORBATE AS A RADIOPROTECTOR OF NORMAL TISSUE

Beyond the cytotoxicity of P-AscH$^-$ to various cancer cell lines, there is increasing data to suggest that P-AscH$^-$ may be beneficial in providing benefits beyond tumor cytotoxicity by also reversing chemoradiation-induced normal tissue injury. There is growing evidence suggesting that P-AscH$^-$ acts as a pro-oxidant locally on tumor cells but acts as an antioxidant systemically in normal tissues. As previously demonstrated, P-AscH$^-$ generates undetectable levels of H_2O_2 because of low levels of redox active labile metal ions and the presence of high levels of metabolic pathways that rapidly remove H_2O_2 in normal cells [3]. Thus, the reductive capacity of P-AscH$^-$ may predominate and ameliorate the oxidative distress induced by radiation and chemotherapy in normal tissues and cells when compared to cancer cells [21]. Numerous preclinical studies have demonstrated the ability of P-AscH$^-$ to protect from chemoradiotherapy-induced normal tissue toxicity. As previously mentioned, our group demonstrated that P-AscH$^-$ partially ameliorated radiation-induced jejunal crypt loss [17]. Others have shown similar results of inhibition of radiation-induced ileal goblet cell toxicity with systemic P-AscH$^-$ [22], while others have demonstrated that oral ascorbate protected jejunal villi as well as increased mouse survival following total abdominal radiation [23]. Administration of ascorbate significantly reduced levels of ROS in

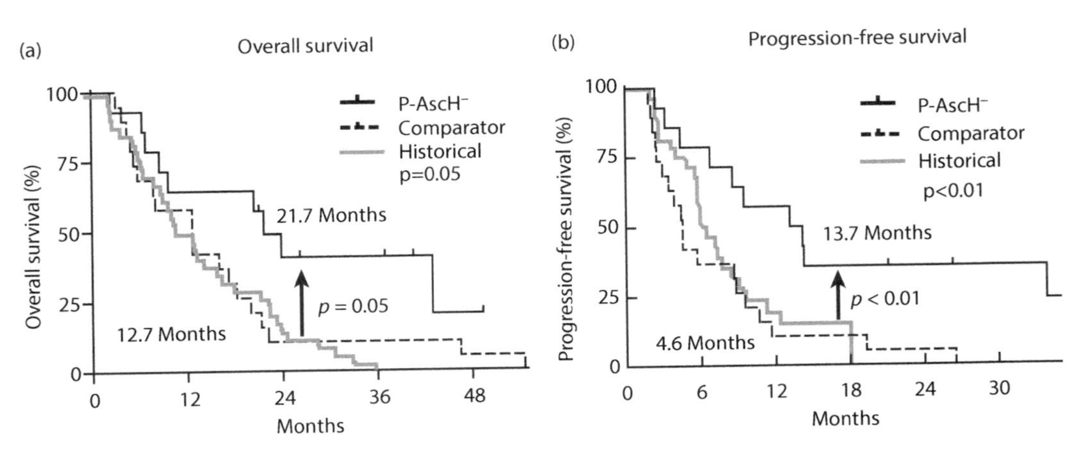

Figure 5.4. Survival. (a) Median OS of subjects treated with P-AscH$^-$, gemcitabine, and radiation was 21.7 months compared to a median OS of 12.7 months in the comparator arm of the study ($p = .05$). This rate is also increased compared to historical median OS of 11.1 months published by Loehrer et al. (b) Median PFS was 13.7 months in the P-AscH$^-$-treated patients and 4.6 months in the comparator patients treated with gemcitabine and radiation ($p < .01$). (OS, overall survival; PFS, progression-free survival.)

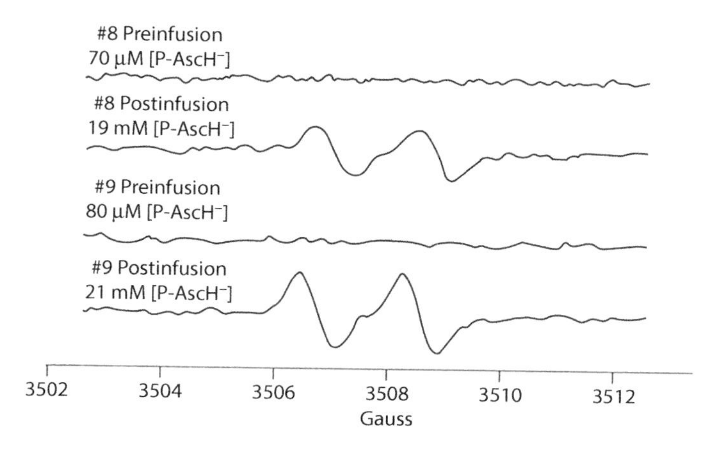

Figure 5.5. Ascorbate radical is observed in whole blood only with high levels of ascorbate. Ascorbate radical is below the limit of detection (<10 nM under these experimental conditions) in preinfusion samples of whole blood that have typical nutritional levels of ascorbate, here 70–80 μM. Ascorbate radical (100–150 nM) is easily detectable in postinfusion samples that have very high levels of ascorbate (19–21 mM). This presence of ascorbate radical indicates the ongoing oxidation of ascorbate in whole blood ascorbate levels.

the small intestine as well as tissue necrosis factor (TNF-α) 1 week following irradiation. Mice that received ascorbate both pre- and postradiation had increased survival compared to those that were given ascorbate postirradiation. Combined, these studies support the hypothesis that the antioxidant properties of ascorbate protect the gastrointestinal tract from radiation-induced toxicity.

The protective effects of P-AscH⁻ may not be limited to the gut. Alopecia and achromotrichia are common side effects of radiation therapy [24]. Daily P-AscH⁻ significantly delayed the onset of radiation-induced achromotrichia and delayed the onset and reduced the incidence of radiation-induced alopecia [25]. Additionally, there is preliminary evidence that P-AscH⁻ may protect red blood cells from osmotic hemolysis caused by oxidative distress induced by the Taxol class of chemotherapeutic agents [26]. Combined, these preclinical studies offer persuasive evidence that P-AscH⁻ may protect normal tissue from chemoradiation while simultaneously sensitizing tumors to the same therapy. However, additional preclinical studies are needed to explore the mechanism of how P-AscH⁻ can lessen normal tissue toxicity associated with radiochemotherapies.

Clinical studies using P-AscH⁻ for a variety of cancers have also indirectly demonstrated the ability of P-AscH⁻ to protect from chemoradiotherapy-induced normal tissue toxicity. A multi-institutional study evaluating quality of life in postsurgical breast cancer patients receiving adjuvant chemoradiation demonstrated that P-AscH⁻ during adjuvant chemoradiation resulted in significantly fewer side effects (i.e., nausea, loss of appetite, fatigue, depression, sleep disorders, and bleeding diathesis) compared to adjuvant chemoradiation alone [27]. In addition, there was a better performance status in patients receiving ascorbate in addition to chemoradiation. Another phase I clinical trial exclusively examined the impact of P-AscH⁻ on quality of life in patients with various other cancers [28]. In this prospective, observational study, 60 patients with various newly diagnosed malignant cancers receiving adjuvant P-AscH⁻ had significantly improved quality of life scores at both 2 and 4 weeks of treatment when compared to initial scores. Specifically, patients receiving P-AscH⁻ had significant increases in physical, role, emotional, cognitive, and social functioning after 4 weeks of intravenous ascorbate therapy, and significant relief of fatigue, pain, insomnia, constipation, and financial difficulties. In addition, physicians who were treating the patients reported that 47% of their patients at 2 weeks and 60% of their patients at 4 weeks had minimally to much improved quality of life. At the conclusion of their study, the authors proposed that P-AscH⁻ can safely improve the quality of life of cancer patients, and they suggest further use of intravenous ascorbate as a palliative care procedure, focusing on improving quality of life and, in particular, relief from fatigue.

P-AscH$^-$ may also affect plasma F_2-isoprostane levels, a marker of systemic oxidative damage caused by lipid peroxidation. Welsh et al. demonstrated significant decreases in F_2-isoprostanes in patients receiving P-AscH$^-$ in combination with gemcitabine chemotherapy for stage IV pancreatic cancer [3]. In another study at the University of Iowa, patients receiving chemoradiation for locally advanced pancreatic cancer who also received P-AscH$^-$ showed a significant decrease over time in F_2-isoprostanes when compared to patients receiving the same chemoradiation but without P-AscH$^-$ [20]. In a phase I/IIa pilot trial of P-AscH$^-$ in combination with carboplatin/paclitaxel chemotherapy in patients with advanced stage ovarian cancer, there were decreased grade III and grade IV therapy-related toxicities in patients who were administered adjuvant P-AscH$^-$ as compared to those in the chemotherapy alone group [29]. Taken together, these clinical studies demonstrate that P-AscH$^-$ may act as an antioxidant in normal cells, while acting as a pro-oxidant in cancer cells.

Specifically for pancreatic cancer, the dose limitation of radiation is dictated by damage to surrounding tissue, in particular, the small intestine [30]. The expected complication rate when the intestine receives a total dose of 50 Gy in 1.8 or 2 Gy fractions is 5% but increases substantially to 50% when radiation doses reach a total dose of 60 Gy [31]. The maximum recommended conventional radiation dose for the treatment of pancreatic adenocarcinoma is 54 Gy, which poses a 9.4%–24.4% complication rate [30,32]. Unfortunately, efforts to find radioprotection agents that could potentially allow increased radiation dosing while avoiding concurrent tumor protection have been relatively unsuccessful [33]. Recently, our group has demonstrated that P-AscH$^-$ selectively radioprotects normal cells relative to cancer cells *in vitro* and *in vivo*. P-AscH$^-$ reverses radiation-induced clonogenic cell death in nontumorigenic cells but radiosensitized a variety of human pancreatic cancer cell lines [20].

The mechanisms underlying the differential effects of P-AscH$^-$ in radiation-induced DNA damage in cancer versus normal cells are unknown at this time. The clinical effect of radiation damage on the small intestine is demonstrated by villous blunting, loss of crypt cells, and collagen deposition, which was partially ameliorated with P-AscH$^-$ [20]. Jejunum from mice treated with radiation and P-AscH$^-$ displayed lower levels of glutathione disulfide (GSSG) suggesting that P-AscH$^-$ may be acting as a systemic antioxidant to normal tissue during radiation. In addition, 4-hydroxy-2-nonenal-(4HNE) modified protein expression, a marker of protein oxidation, demonstrated that irradiated jejunal tissue treated with P-AscH$^-$ expressed lower levels of 4HNE than irradiated tumor tissue treated with P-AscH$^-$. Taken together, these data suggest that P-AscH$^-$ radiosensitizes pancreatic tumors and acts as a pro-oxidant in irradiated tumor tissue but may act as an antioxidant in radiated normal tissue. Another possible explanation is that the H_2O_2 produced by P-AscH$^-$ has a much greater detrimental effect on cancer cells than normal cells due to the fact that cancer cells have elevated levels of reactive metals and are likely to have increased formation of hydroxyl radical, causing DNA damage [34,35]. Normal cells, which have relatively lower fluxes of superoxide ($O_2^{\bullet-}$) as seen in Figure 5.3, have lower levels of labile metals, and more abundant catalase to detoxify H_2O_2, leaving P-AscH$^-$ to act as an antioxidant alone [36,37]. Others have demonstrated that the NF-κB transcription factor RelB can determine the differential effects of P-AscH$^-$ in normal versus cancer cells [38].

SUMMARY

For many patients with cancer, adjuvant chemotherapy and radiotherapy remain the only treatment option with the goal of palliating symptoms and improving quality of life. Ionizing radiation has been shown to increase the levels of catalytic iron in tissues, which increases the rate of ascorbate-induced peroxide formation and cytotoxicity [36,39]. One of the major factors limiting the implementation of radiosensitizers and radioprotectors in clinical cancer care has been the nonspecific mechanism of action of many radiation modulators. Drugs that affect radiosensitivity by targeting key cell survival pathways are likely to affect both tumors and normal tissues. Thus, selectivity is an important goal for designing improved radiation modulators. P-AscH$^-$ appears to be a promising agent in this regard. Therefore, more clinical investigations including phase II and phase III trials to determine whether P-AscH$^-$ acts synergistically with radiation therapy as a radiosensitizer to increase radiation-induced cancer cell death are needed.

ACKNOWLEDGEMENTS

Supported by National Institutes of Health (NIH) grants CA184051, P01 CA217797, CA148062, and the Medical Research Service, Department of Veterans Affairs, 1I01BX001318-01A2.

REFERENCES

1. Siegel, R. L., Miller, K. D. and Jemal, A. 2018 Cancer statistics, 2018. *CA Cancer J Clin.* 68, 7–30.
2. Iacobuzio-Donahue, C. A., Fu, B. J., Yachida, S., Luo, M. D., Abe, H., Henderson, C. M., Vilardell, F. et al. 2009 DPC4 gene status of the primary carcinoma correlates with patterns of failure in patients with pancreatic cancer. *J. Clin. Oncol.* 27, 1806–1813.
3. Welsh, J. L., Wagner, B. A., van't Erve, T. J., Zehr, P. S., Berg, D. J., Halfdanarson, T. R., Yee, N. S. et al. 2013 Pharmacological ascorbate with gemcitabine for the control of metastatic and node-positive pancreatic cancer (PACMAN): Results from a phase I clinical trial. *Cancer Chemother. Pharmacol.* 71, 765–775.
4. Levine, M., Conry-Cantilena, C., Wang, Y., Welch, R. W., Washko, P. W., Dhariwal, K. R., Park, J. B. et al. 1996 Vitamin C pharmacokinetics in healthy volunteers: Evidence for a recommended dietary allowance. *Proc. Natl. Acad. Sci. USA* 93, 3704–3709.
5. Du, J., Martin, S. M., Levine, M., Wagner, B. A., Buettner, G. R., Wang, S. H., Taghiyev, A. F., Du, C., Knudson, C. M. and Cullen, J. J. 2010 Mechanisms of ascorbate-induced cytotoxicity in pancreatic cancer. *Clin. Cancer Res.* 16, 509–520.
6. Chen, Q., Espey, M. G., Krishna, M. C., Mitchell, J. B., Corpe, C. P., Buettner, G. R., Shacter, E. and Levine, M. 2005 Pharmacologic ascorbic acid concentrations selectively kill cancer cells: Action as a pro-drug to deliver hydrogen peroxide to tissues. *Proc. Natl. Acad. Sci. USA* 102, 13604–13609.
7. Chen, Q., Espey, M. G., Sun, A. Y., Lee, J. H., Krishna, M. C., Shacter, E., Choyke, P. L., Pooput, C., Kirk, K. L., Buettner, G. R. and Levine, M., 2007 Ascorbate in pharmacologic concentrations selectively generates ascorbate radical and hydrogen peroxide in extracellular fluid in vivo. *Proc. Natl. Acad. Sci. USA* 104, 8749–8754.
8. Wagner, B. A., Witmer, J. R., van 't Erve, T. J. and Buettner, G. R. 2013 An assay for the rate of removal of extracellular hydrogen peroxide by cells. *Redox. Biol.* 1, 210–217.
9. Cieslak, J. A., Strother, R. K., Rawal, M., Du, J., Doskey, C. M., Schroeder, S. R., Button, A., Wagner, B. A., Buettner, G. R. and Cullen, J. J. 2015 Manganoporphyrins and ascorbate enhance gemcitabine cytotoxicity in pancreatic cancer. *Free Radic. Biol. Med.* 83, 227–237.
10. Espey, M. G., Chen, P., Chalmers, B., Drisko, J., Sun, A. Y., Levine, M. and Chen, Q. 2011 Pharmacologic ascorbate synergizes with gemcitabine in preclinical models of pancreatic cancer. *Free Radic. Biol. Med.* 50, 1610–1619.
11. Chen, Y., Jungsuwadee, P., Vore, M., Butterfield, D. A. and St Clair, D. K. 2007 Collateral damage in cancer chemotherapy: Oxidative stress in nontargeted tissues. *Mol. Interv.* 7, 147–156.
12. Rawal, M., Schroeder, S. R., Wagner, B. A., Cushing, C. M., Welsh, J. L., Button, A. M., Du, J., Sibenaller, Z. A., Buettner, G. R. and Cullen, J. J. 2013 Manganoporphyrins increase ascorbate-induced cytotoxicity by enhancing H2O2 generation. *Cancer Res.* 73, 5232–5241.
13. Du, J., Nelson, E. S., Simons, A. L., Olney, K. E., Moser, J. C., Schrock, H. E., Wagner, B. A. et al. 2013 Regulation of pancreatic cancer growth by superoxide. *Mol. Carcinog.* 52, 555–567.
14. Davies, K. J. 1999 The broad spectrum of responses to oxidants in proliferating cells: A new paradigm for oxidative stress. *IUBMB Life.* 48, 41-47.
15. Ridnour, L. A., Oberley, T. D. and Oberley, L. W. 2004 Tumor suppressive effects of MnSOD overexpression may involve imbalance in peroxide generation versus peroxide removal. *Antioxid. Redox. Signal.* 6, 501–512.
16. Schafer, F. Q. and Buettner, G. R. 2001 Redox environment of the cell as viewed through the redox state of the glutathione disulfide/glutathione couple. *Free Radic. Biol. Med.* 30, 1191–1212.
17. Du, J., Cieslak, J. A., 3rd, Welsh, J. L., Sibenaller, Z. A., Allen, B. G., Wagner, B. A., Kalen, A. L. et al. 2015 Pharmacological ascorbate radiosensitizes pancreatic cancer. *Cancer Res.* 75, 3314–3326.
18. Loehrer, P. J., Sr., Feng, Y., Cardenes, H., Wagner, L., Brell, J. M., Cella, D., Flynn, P. et al. 2011 Gemcitabine alone versus gemcitabine plus radiotherapy in patients with locally advanced pancreatic cancer: An Eastern Cooperative Oncology Group trial. *J. Clin. Oncol.* 29, 4105–4112.
19. Storer, B. E. 1989 Design and analysis of phase I clinical trials. *Biometrics.* 45, 925–937.
20. Alexander, M. S., Wilkes, J. G., Schroeder, S. R., Buettner, G. R., Wagner, B. A., Du, J., Gibson-Corley, K. et al. 2018 Pharmacologic ascorbate

reduces radiation-induced normal tissue toxicity and enhances tumor radiosensitization in pancreatic cancer. *Cancer Res.* 78, 6838–6851.

21. Conklin, K. A. 2004 Chemotherapy-associated oxidative stress: Impact on chemotherapeutic effectiveness. *Integr. Cancer Ther.* 3, 294–300.
22. Kanter, M. and Akpolat, M. 2008 Vitamin C protects against ionizing radiation damage to goblet cells of the ileum in rats. *Acta. Histochem.* 110, 481–490.
23. Ito, Y., Kinoshita, M., Yamamoto, T., Sato, T., Obara, T., Saitoh, D., Seki, S. and Takahashi, Y. 2013 A combination of pre- and post-exposure ascorbic acid rescues mice from radiation-induced lethal gastrointestinal damage. *Int. J. Mol Sci.* 14, 19618–19635.
24. Freites-Martinez, A., Shapiro, J., Goldfarb, S., Nangia, J., Jimenez, J. J., Paus, R. and Lacouture, M. E. 2018 CME part 1: Hair disorders in cancer patients. *J. Am. Acad. Dermatol.* 80.
25. Schoenfeld, J. D., Alexander, M. S., Waldron, T. J., Sibenaller, Z. A., Spitz, D. R., Buettner, G. R., Allen, B. G. and Cullen, J. J. 2019 Pharmacological ascorbate as a means of sensitizing cancer cells to radio chemotherapy while protecting normal tissue. *Semin. Radiat. Oncol.* 29, 25–32.
26. Alatawi, F., Faridi, U. and Mostafa, M. 2017 Protective role of tocopherol and ascorbic acid in taxol-treated human erythrocytes in vitro. *Toxicol. Res. Appl.* 1, 2397847317705813.
27. Vollbracht, C., Schneider, B., Leendert, V., Weiss, G., Auerbach, L. and Beuth, J. 2011 Intravenous vitamin C administration improves quality of life in breast cancer patients during chemo-/radiotherapy and aftercare: Results of a retrospective, multicentre, epidemiological cohort study in Germany. *In Vivo.* 25, 983–990.
28. Takahashi HM, H. M. and Yanagisawa, A. 2012 High-dose intravenous vitamin C improves quality of life in cancer patients. *Personalized Medicine Universe.* 1, 49–53.
29. Ma, Y., Chapman, J., Levine, M., Polireddy, K., Drisko, J. and Chen, Q. 2014 High-dose parenteral ascorbate enhanced chemosensitivity of ovarian cancer and reduced toxicity of chemotherapy. *Sci. Transl. Med.* 6, 222ra218.
30. Bahl, A., Bhattacharyya, T., Kapoor, R., Singh, O. A., Parsee, T. and Sharma, S. C. 2013 Postoperative radiotherapy in periampullary cancers: A brief review. *J. Gastrointest. Cancer* 44, 111–114.
31. Perrakis, N., Athanassiou, E., Vamvakopoulou, D., Kyriazi, M., Kappos, H., Vamvakopoulos, N. C. and Nomikos, I. 2011 Practical approaches to effective management of intestinal radiation injury: Benefit of resectional surgery. *World J. Gastroenterol.* 17, 4013–4016.
32. Huguet, F., Goodman, K. A., Azria, D., Racadot, S. and Abrams, R. A. 2012 Radiotherapy technical considerations in the management of locally advanced pancreatic cancer: American-French consensus recommendations. *Int. J. Radiat. Oncol. Biol. Phys.* 83, 1355–1364.
33. Moding, E. J., Kastan, M. B. and Kirsch, D. G. 2013 Strategies for optimizing the response of cancer and normal tissues to radiation. *Nat. Rev. Drug. Discov.* 12, 526–542.
34. Schoenfeld, J. D., Sibenaller, Z. A., Mapuskar, K. A., Wagner, B. A., Cramer-Morales, K. L., Furqan, M., Sandhu, S. et al. 2017 O2(-) and H2O2-mediated disruption of Fe metabolism causes the differential susceptibility of NSCLC and GBM cancer cells to pharmacological ascorbate. *Cancer Cell.* 31, 487–500 e488.
35. Du, J., Wagner, B. A., Buettner, G. R. and Cullen, J. J. 2015 Role of labile iron in the toxicity of pharmacological ascorbate. *Free Radical. Bio. Med.* 84, 289–295.
36. Buettner, G. R. and Jurkiewicz, B. A. 1996 Catalytic metals, ascorbate and free radicals: Combinations to avoid. *Radiat. Res.* 145, 532–541.
37. Doskey, C. M., Buranasudja, V., Wagner, B. A., Wilkes, J. G., Du, J., Cullen, J. J. and Buettner, G. R. 2016 Tumor cells have decreased ability to metabolize H_2O_2: Implications for pharmacological ascorbate in cancer therapy. *Redox. Biol.* 10, 274–284.
38. Wei, X., Xu, Y., Xu, F. F., Chaiswing, L., Schnell, D., Noel, T., Wang, C., Chen, J., St Clair, D. K. and St Clair, W. H. 2017 RelB expression determines the differential effects of ascorbic acid in normal and cancer cells. *Cancer Res.* 77, 1345–1356.
39. Persson, H. L. 2006 Radiation-induced lysosomal iron reactivity: Implications for radioprotective therapy. *IUBMB Life.* 58, 395–401.

CHAPTER SIX

Vitamin C and Somatic Cell Reprogramming

Jingjing Wang and Luisa Cimmino

CONTENTS

INTRODUCTION

The potential to reprogram somatic cells into induced pluripotent stem cells (iPSCs), with the capacity to generate embryos and all types of differentiated cell lineages, has provided an invaluable tool in the study of development and disease, and in generating cells with novel therapeutic applications in regenerative medicine [1,2]. The enforced expression of a defined set of transcription factors such as Oct4, Sox2, Klf4, and c-Myc (OSKM) is sufficient to generate iPSCs from somatic cells that replicate the naïve "ground state" of blastocyst-derived embryonic stem cells (ESCs) [1]. However, reprogramming efficiency is often very low and influenced by multiple factors including donor cell age, passages in culture, lineage of origin, or stage of development [3–5]. Embryonic tissues are the most efficiently reprogrammed, producing iPSCs that are nearly identical to ESCs from fertilized embryos [2]. When comparing adult donor cells, keratinocytes and blood cells reprogram more efficiently than fibroblasts [6,7], and immature progenitors reprogram with greater efficiency than terminally differentiated blood cells [5]. Genetic and epigenetic abnormalities can persist in iPSCs, such as copy number variations [8], point mutations [9], dysregulated expression of imprinted genes, and aberrant DNA methylation patterns that may limit both reprogramming efficiency and iPSC therapeutic potential [10–12].

Reprogrammed cells can also retain an epigenetic memory of their tissue of origin that favors their differentiation toward the donor cell lineage, imparting a cell intrinsic bias that may restrict their capacity for widespread use in disease modeling or treatment [7,13].

Studies over the last decade have revealed that the addition of vitamin C (ascorbic acid, ascorbate) to the culture media of somatic cells during reprogramming vastly improves both the efficiency and quality of iPSC formation [14–16]. Vitamin C, in addition to its role as a cellular antioxidant, participates as a critical cofactor of Fe^{2+} and α-ketoglutarate-dependent dioxygenases (α-KGDDs) to enhance their catalytic activity (reviewed in [17]). Included among the diverse family of α-KGDDs are epigenetic regulators such as the Jumonji-C (JmjC) domain-containing histone demethylases (JHDMs), and the ten-eleven translocation (TET) family of DNA hydroxylases that regulate histone demethylation and DNA hydroxymethylation, respectively, to drive the epigenetic reprogramming of somatic cells into pluripotent stem cells [15,16,18,19].

Vitamin C can also reprogram the epigenome of adult stem cells. Epigenetic dysregulation is a hallmark of hematopoietic stem cell (HSC) transformation, and impaired TET function is a driver of blood cell malignancies [20,21]. By enhancing TET function, vitamin C has been shown to block aberrant self-renewal and slow down leukemia progression, highlighting its potential therapeutic benefit for the treatment of patients with hematopoietic malignancies [22,23]. The ability of vitamin C to directly modulate the epigenome has expanded our appreciation of this essential vitamin in the regulation of cell fate that has widespread application in the fields of stem cell biology and regenerative medicine and in the treatment of cancer.

EPIGENETIC REPROGRAMMING OF SOMATIC CELLS BY VITAMIN C

The epigenetic state of a cell is determined by a unique pattern of histone modifications and DNA methylation that directs lineage-specific gene expression patterns. The epigenetic memory of these expression patterns is established during development and differentiation and must be erased from somatic cells in order to reset the genome of iPSCs to an ESC-like state.

Early in 2010, Esteban and colleagues were the first to report that vitamin C supplementation could increase the reprogramming efficiency of both human and mouse somatic cells into iPSCs, and convert pre-iPSC colonies to fully potent iPSCs [15]. Vitamin C increases the rate of ESC proliferation [24] and the speed of reprogramming [15] and was originally added to the culture media for its antioxidant properties in an attempt to counteract the reactive oxygen species (ROS) generated during reprogramming [15]. However, in comparison to other antioxidants such as glutathione, N-acetylcysteine, vitamin E, and lipoic acid, vitamin C was found to be substantially more efficient at enhancing iPSC generation [18,19]. Moreover, co-culture with inhibitors of the vitamin C–dependent α-KGDDs, such as the iron chelator desferrioxamine (DFO) or the α-KG analog dimethyloxalylglycine (DMOG), led to impaired iPSC formation [16], thereby implicating these enzymes in the mechanism of vitamin C–mediated somatic cell reprogramming. Histone demethylases at the time were known to be important for the expression of the ESC master transcription factor Nanog [25], and based on the role of vitamin C as a cofactor for α-KGDDs, such as JHDMs, it was postulated that the mechanism by which vitamin C could facilitate reprogramming was through increased histone demethylation [15].

A subsequent independent study found that vitamin C could promote DNA demethylation in ESCs at genomic loci known to undergo widespread loss of methylation during the reprogramming of somatic cells into iPSCs [24]. Furthermore, Stadtfeld and colleagues [14] showed that vitamin C prevents DNA hypermethylation and maintains the active chromatin state and normal maternal expression of the paternally imprinted *Dlk1-Dio3* gene cluster. Loss of imprinting at this locus due to DNA hypermethylation was known to cause the abnormal development of mice generated from iPSCs [10]. The ability of vitamin C to increase the efficiency of reprogramming and improve the quality of iPSCs has therefore been attributed to its ability to modulate the epigenome (Figure 6.1a).

Vitamin C Enhances Histone Demethylation during Reprogramming

Histone demethylases were initially proposed to be the key effectors of iPSC reprogramming in response to vitamin C treatment. Vitamin C is required for the optimal activity and demethylation capacity of several JHDMs [26], which include over 20 α-KGDDs in humans that hydroxylate

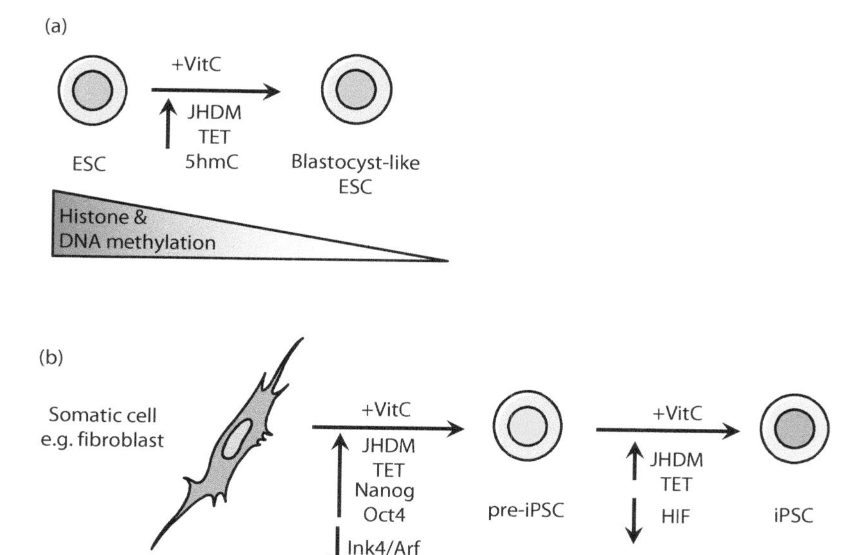

Figure 6.1. Vitamin C enhances the efficiency and quality of somatic cell reprogramming. (a) Vitamin C treatment enhances the activity of epigenetic regulators in embryonic stem cells (ESCs), including jumonji histone demethylase (JHDMs) and ten-eleven translocation (TET) proteins, leading to hypomethylation of histones and DNA to establish the blastocyst-like state of embryo-derived ESCs. (b) The addition of vitamin C to the culture of somatic cells during reprogramming increases the expression of pluripotency genes (Nanog and Oct4) and suppresses the expression of senescence-inducing factors (Ink4/Arf). Vitamin C may enhance reprogramming fidelity by regulating other Fe^{2+} and α-ketoglutarate-dependent dioxygenases, such as hypoxia-inducible factor (HIF) prolyl hydroxylases, in addition to JHDMs and TETs.

and remove mono-, di-, or trimethyl-lysine marks in histones [27]. Histone demethylation is catalyzed by the JmjC domain to produce a highly reactive oxoferryl species that hydroxylates the methylated substrate, allowing spontaneous loss of the methyl group as formaldehyde [28]. Targeting lysines in the core histone H3, JHDM1 (Jhdm1a/b, KDM2) specifically demethylates H3K36; JHDM2A (JMJD1A, KDM3A) demethylates H3K9; JHDM3A (JMJD2A, KDM4) demethylates both trimethylated H3K9 and H3K36, and UTX (KDM6A) demethylates H3K27, all of which can be modulated by vitamin C to regulate chromatin state and gene expression [25,27,29,30].

H3K9 methylation has been shown to be a barrier to somatic cell reprogramming [31]. Vitamin C, by increasing JMJD1 (KDM3) activity, induces a specific loss of H3K9me2 in ESCs [32]. Vitamin C can also drive pre-iPSC-to-iPSC transition by enhancing both JMJD1 (KDM3) and JMJD3 (KDM4)-mediated H3K9me2/3 demethylation at core pluripotency gene loci [31]. UTX (KDM6A) is another JmjC-domain-containing α-KGDD that demethylates H3K27 and is a crucial regulator of pluripotency induction during somatic cell reprogramming [33]. Vitamin C–treated ESCs in culture change their distribution of H3K27me3 across the genome [32]; however, the effects on total levels are minor, and further studies into the reprogramming stage or tissue-specific role for vitamin C in modulating UTX function are required. Vitamin C can also increase the activity of JHDM1a/1b to promote H3K36me2/3 demethylation in mouse embryonic fibroblasts in culture and during iPSC reprogramming [16]. Consistent with these observations, overexpression of *Jhdm1a/1b* potently enhances reprogramming, whereas knockdown impairs iPSC generation [16].

Vitamin C also enhances iPSC generation at least in part by delaying cell senescence [15]. Reprogramming is proliferation dependent to permit the time required for the progressive acquisition of a pluripotent state [34]. The *Cdkn2a* (*Ink4/Arf*) locus is a critical regulator of senescence that causes growth arrest in cells, and expression of these cell cycle regulators represents a roadblock to reprogramming [4]. Aged cells exhibit higher levels of Ink4/Arf, which limits their efficiency and fidelity of reprogramming [4]. The histone demethylase JHDM1b enhances fibroblast proliferation in part through the repression of *Ink4/Arf*, by removing H3K36me2/3 marks that are associated with active expression from this locus [35,36]. The JHDM1b-mediated removal of H3K36me2/3 is enhanced by vitamin C, increasing the efficiency of silencing of the *Ink4/Arf* locus in order to overcome this senescence roadblock [16]. Active silencing of the *Ink4/Arf* locus in the presence of vitamin C could also help iPSCs derived from aged donor cells to reprogram more efficiently.

Vitamin C Enhances DNA Demethylation during Reprogramming

Enhanced iPSC reprogramming and DNA demethylation induced by vitamin C are also mediated by the TET proteins [18,19]. The TETs (TET1-3) are a subfamily of α-KGDDs that catalyze the hydroxylation of 5-methylcytosine (5 mC) residues in DNA to generate 5-hydroxymethylcytosine (5 hmC) [37], and through successive oxidation reactions, 5-formylcytosine (5 fC) and 5-carboxylcytosine (5 CaC). The oxidative products of 5 mC catalyzed by TET proteins can be stable modifications in the genome or transient modifications that provide a trigger for active or passive DNA demethylation [37–39].

TET1 and TET2 are the most highly expressed TET proteins in embryonic stem cells (ESCs) and together are responsible for ~80% of the 5 hmC observed in the ESC genome [40,41]. OSKM-mediated reprogramming of mouse embryonic fibroblasts (MEFs) activates *Tet2* expression and induces a genome-wide increase in 5 hmC and DNA demethylation at the loci of pluripotency factors such as *Nanog* and *Esrrb* leading to their enhanced expression [42] (Figure 6.1b). TET1 and TET2 proteins also physically interact with NANOG, and their overexpression promotes hydroxylation and DNA demethylation at the *Oct4* locus, leading to increased reprogramming efficiency [43,44]. Furthermore, TET expression is essential for reprogramming, given that *Tet1*- or *Tet2*-depleted MEFs are unable to generate OSKM-mediated iPSCs [42,43].

Vitamin C can dramatically increase 5 hmC production in ESCs and during the reprogramming of mouse and human fibroblasts to iPSCs [18,45,46] (Figure 6.1b). Notably, 100 μM vitamin C is sufficient to increase 5 hmC by up to approximately fourfold above basal levels in ESCs within 24 hours of treatment, with even larger effects on the levels of 5 fC (10-fold increase) and 5 caC (20-fold increase) [19]. All three TET proteins bind preferentially at transcriptional start sites (TSSs) and promoters with affinities that positively correlate with CpG density [45,47–51]. Consistent with TET binding profiles, the rapid increase in 5 hmC observed in the genome of ESCs treated with vitamin C accumulates at TSSs and is followed by DNA demethylation at the promoters of germ line genes normally expressed during formation of a blastocyst-like state [18,19].

TET1-3 triple knockout ESCs cannot support embryonic development, and their combined loss abolishes all 5 hmC in ESCs leading to impaired differentiation during ESC lineage specification and global DNA hypermethylation [52]. The effect of vitamin C to enhance 5 hmC, gene expression, and DNA demethylation is also lost in $Tet1^{-/-}Tet2^{-/-}$ double-knockout ESCs [18,19], highlighting the importance of TET proteins as downstream targets of vitamin C–mediated DNA demethylation. Aberrant imprinting marks in the *Dlk1* locus have been reported in the progeny of mice carrying mutations in both Tet1 and Tet2 [41], and vitamin C prevents the aberrant DNA hypermethylation and silencing of imprinted genes at the *Dlk1-Dio3* gene cluster [14], suggesting that a combination of increased JHDM and TET activity regulates this locus to enhance reprogramming.

Vitamin C can also work synergistically with other factors, such as vitamin A (retinol, retinoic acid), to enhance iPSC reprogramming [15,53,54]. Retinoic acid (RA) can reduce the effective concentration of vitamin C required for efficient reprogramming of serum-grown epiblast stem cells (EpiSCs) to naïve pluripotent ESCs [54]. RA has no direct effect on TET enzymatic activity but can increase TET2 and TET3 expression, leading to increased 5 hmC and enhanced DNA demethylation [54]. When supplemented in combination, RA signaling increases TET expression levels, and vitamin C in turn will enhance the activity of an increased pool of TET protein, resulting in a synergistic increase in reprogramming efficiency. Although TETs function in general as positive regulators of reprogramming [43,44], further studies have revealed that more TET activity might not always be better. Using MEFs as a model system, it has been reported that in the absence of vitamin C, both Tet1 and Tet2 promote iPSC generation; but with increasing concentrations of vitamin C, overexpressed Tet1 gradually becomes an inhibitor of reprogramming, while Tet2 remains an enhancer [55]. TET1 is known to play a dual role in transcription, activating or repressing its various target genes [47–49], whereas TET2 binding at promoters positively correlates with gene expression [45]. Taken together, these data suggest that the correct balance in activity between the TET proteins is integral to reprogramming. Further studies may be required to determine the optimal amount of vitamin C, in combination with other factors such as RA, that should be used to reprogram adult or embryonic tissues for the generation of iPSCs.

ADDITIONAL ROLES FOR VITAMIN C IN SOMATIC CELL REPROGRAMMING

The most important known role of vitamin C as a modulator of somatic cell reprogramming is through its ability to directly influence the activity of α-KGDDs. In addition to the epigenetic changes driven by JHDMs and TETs, vitamin C can target other α-KGDDs that regulate metabolism, DNA repair, and DNA/RNA demethylation in reprogramming cells.

Hypoxia-Inducible Factors and Somatic Cell Reprogramming

Hypoxia-inducible factors (HIFs) are oxygen-sensing transcription factors whose stability can be regulated by prolyl hydroxylases, a subfamily of α-KGDDs [56]. HIFs are made up of α and β subunits that dimerize and translocate to the nucleus in response to hypoxia to regulate the expression of genes involved in oxygen homeostasis, glucose metabolism, angiogenesis, erythropoiesis, and iron metabolism [57,58]. Hypoxia is a key feature of the stem cell niche, known to increase the self-renewal capacity of ESCs, adult stem cells, and enhance the generation of iPSCs [57,59]. During reprogramming, the inhibition of glycolysis reduces reprogramming efficiency, whereas stimulation of glycolytic activity enhances iPSC generation [60,61], and the HIF proteins are required to initiate the metabolic switch from oxidative to glycolytic metabolism [62,63]. In conditions of hypoxia, the HIFα subunits (HIF1α and HIF2α) are stabilized and help drive the metabolic switch to glycolysis, an essential step for the initial stages of somatic cell reprogramming; however, prolonged stabilization of HIF-2α in the final stages of reprogramming will cause a significant block in the acquisition of a fully pluripotent ESC-like state [63]. Vitamin C treatment mimics conditions of normoxia, enhancing the activity of HIF prolyl hydroxylases, which hydroxylate the HIF proteins [56] and lead to the binding of an E3 ubiquitin ligase, polyubiquitination, and proteasomal degradation [56]. Vitamin C also enhances the activity of the asparaginyl hydroxylase factor inhibiting HIF-1 (FIH-1), an important suppressor of the transcriptional activity of HIF [64] that correlates with reports of vitamin C reducing the mRNA expression levels of HIF genes in leukemia cell lines [65]. Vitamin C may also have stage-specific roles in fine-tuning the process of reprogramming. When pre-iPSCs, which are trapped at an intermediate state of reprogramming, are treated with vitamin C, conversion to fully reprogrammed iPSCs is induced [6]. Given that HIF-2α suppresses reprogramming during the latter stages of iPSC generation, prolyl hydroxylase-mediated degradation of HIF-2α may be the mechanism by which vitamin C increases the efficiency of reprogramming at this crucial final step.

ALKB Homologues and Somatic Cell Reprogramming

The alkylated DNA repair protein AlkB homologues (ALKBHs) are another group of α-KGDDs that can influence somatic cell reprogramming and be targeted for enhanced activity by vitamin C. ALKBH1 is a histone dioxygenase that removes methyl groups from histone H2A to regulate gene expression [66,67]. ALKBH1 expression is higher in stem cells than differentiated cells and increases during iPSC reprogramming [68,69]. Moreover, ALKBH1 interacts directly with the core pluripotency factors Oct4, Sox2, and Nanog at overlapping sites on chromatin and influences the expression of miRNAs important for maintaining ESC self-renewal and pluripotency [67]. Other AlkB homologues are indirectly implicated in the regulation of somatic cell reprogramming. RNA methyltransferases generate m^6A modifications in mRNA transcripts that can be removed by two RNA demethylases—the AlkB homologues known as the fat-mass and obesity-associated (*FTO*) gene and AlkB homolog 5 (*ALKBH5*) [70,71]. Increased

m^6A abundance promotes the reprogramming of MEFs to pluripotent stem cells; conversely, reduced m^6A levels impede reprogramming [72]. Vitamin C has recently been shown to promote erasure of m^6A during the differentiation of pig oocytes [73]; however, the role of vitamin C in regulating m^6A levels via the modulation of FTO or ALKBH activity during reprogramming has not yet been investigated.

ROLE OF VITAMIN C IN STEM CELL THERAPY AND TISSUE REGENERATION

Stem cell therapies are highly sought after for the rejuvenation and regeneration of tissue in the treatment of aging, neurodegenerative diseases, heart failure, skin or eye disorders, and cancer [11,74]. In the first described clinical trial using iPSCs, skin cells were used to create sheets of retinal pigment epithelium for the treatment of age-related macular degeneration, which effectively stopped further degeneration and brightened a patient's vision [75]. However, due to concerns regarding genetic changes in patient-derived iPSCs, the trial was halted in further patients. The challenges facing the safe and effective implementation of stem cell therapies has not stopped intense research into the development of suitable methods for making relevant cell types for clinical application at high quantity and purity [11,74]. Generating iPSC-based disease models from patients promises to serve both as a renewable source of genetically matched tissues for autologous cell replacement therapy and as an *in vitro* platform for high-throughput screening of small molecules for drug discovery. Limitations to the use of ESCs and iPSCs in stem cell therapies center on their capacity for cell lineage-specific differentiation, which can be more readily achieved for certain cell types, such as neurons, blood, and cardiomyocytes, compared to others [76]. Exogenously supplied vitamin C, in addition to enhancing somatic cell reprogramming, has been shown in numerous studies to increase the rate of proliferation and self-renewal capacity of ESCs, iPSCs, neural stem cells (NSCs), mesenchymal stem cells (MSCs), epithelial or cardiac progenitors, intestinal stem cells, and more (Figure 6.2) (reviewed in [77]) and may serve as a key adjuvant in preclinical models of iPSC-based regenerative medicine.

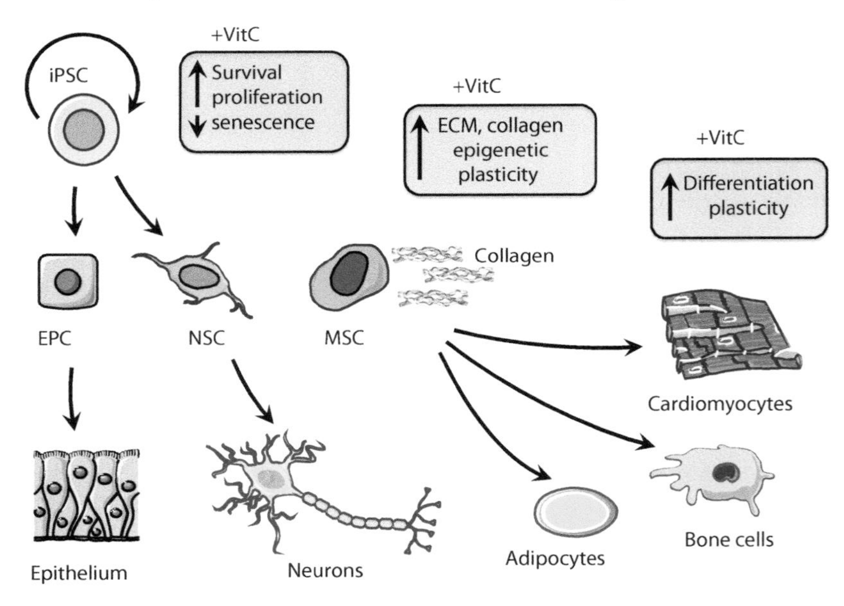

Figure 6.2. Vitamin C as an adjuvant for stem cell therapies and tissue regeneration. Vitamin C has been shown in numerous studies to increase the rate of proliferation and self-renewal capacity of iPSCs and adult stem cells in culture, including epithelial stem and progenitor cells (EPCs), neural stem cells (NSCs), mesenchymal stem cells (MSCs), and more. Vitamin C may serve as a key adjuvant in preclinical models of iPSC-based regenerative medicine. Patient-derived iPSCs can be used as a renewable source of cells for tissue regeneration with potential to form epithelium, neurons, fat tissue (adipocytes), bone, and heart muscle (cardiomyocytes). Vitamin C can maintain stemness of *in vitro* cultures by blocking senescence, improve engraftment by increasing the activity of prolyl hydroxylases to secrete collagen for extracellular matrix (ECM), and maintain epigenetic plasticity to maximize progenitor cell differentiation capacity.

Vitamin C and Neuronal Stem Cells

Neural stem/precursor cells (NSCs) are considered a potential cell source for the treatment of Parkinson disease (PD), which is characterized by the progressive degeneration of dopamine (DA) neurons in the midbrain [78]. Fibroblast-derived iPSCs can be efficiently differentiated into NSCs, giving rise to neuronal and glial cell types in culture that were shown to improve the clinical symptoms of a rat model of Parkinson disease [79]. Similar to the initial rationale for adding vitamin C to somatic cells during reprogramming, several antioxidants were tested in an attempt to mitigate the effect of cellular aging and senescence potentially caused by ROS generation in NSC cultures [80]. Compared to numerous antioxidants tested, only vitamin C treatment protected *in vitro*–expanded NSCs from losing DA neurogenic potential, and transplantation of vitamin C–treated NSCs resulted in improved behavioral restoration, along with enriched DA neuron engraftment in a rat model of PD [80]. The effect on NSC function was attributed to the role of vitamin C as a regulator of DNA demethylation rather than its antioxidant function, and the faithful expression of midbrain-specific markers in engrafted neurons [80]. Another study also described the role of vitamin C in enhancing NSC differentiation toward DA neurons through boosting of Tet1 and Jmjd3 (KDM6B) activity [81]. Treatment with vitamin C may therefore be important for the maintenance of an epigenetic state that favors enhanced survival of cultured NSCs primed for DA neuron differentiation and may improve the efficacy of stem cell therapies generated from iPSCs for the treatment of PD.

Vitamin C and Mesenchymal Stem Cells

Mesenchymal stem cells (MSCs) are multipotent adult stem cells that can differentiate into mesodermal tissues such as bone (osteoblasts), cartilage (chondrocytes), muscle (myocytes), and fat (adipocytes) (reviewed in [82]). MSCs directly harvested from adult organs or generated *in vitro* from iPSCs are highly sought after for use in the field of tissue engineering given their potential to generate a variety of cell types [83–85]; however, ensuring functional regenerative tissue grafting *in vivo* is still a challenge. Compared to scaffold-based approaches, growing MSCs as continuous cell sheets in culture preserves cellular junctions and endogenous extracellular matrix (ECM) that is believed to more readily mimic the organ tissue microenvironment and could improve the quality of tissue transplantation and efficacy of engraftment [82].

Collagens are synthesized and secreted by cells into the extracellular space and are the most abundant proteins to make up the ECM [77]. Vitamin C plays an essential role in the biosynthesis of collagen from procollagen by enhancing the enzymatic activity of the α-KGDD family of collagen prolyl hydroxylases (P4H/P3H) and procollagen-lysine α-KG 5-dioxygenases (PLODs) [86,87]. Prolyl hydroxylation of procollagen is required for the correct folding and stability of collagen, and in the absence of P4H activity, procollagen accumulates in the ER and is not properly secreted [88,89]. In addition to directly enhancing the catalytic activity of collagen prolyl hydroxylases, vitamin C also promotes the transcription of collagen genes and increases the stability of collagen mRNA in multiple cell types [90–93]. ECM is important for maintaining the stem cell niche and for mediating integrin signals during embryonic development, and type-1 collagen facilitates mESC self-renewal [94–96]. Multiple lines of evidence would therefore suggest that vitamin C treatment *in vitro* could improve the regenerative capacity of stem cells by enhancing collagen synthesis and ECM formation in engineered tissues. Vitamin C has indeed been shown to upregulate the expression of extracellular matrix type-I collagen, fibronectin, integrin β1, and the stem cell markers Oct4, Sox2, and Nanog, and promote intercellular junctions that lead to enhanced sheet formation in both *in vivo* animal models and *in vitro* cultured human MSCs [97]. These studies highlight the nonepigenetic influence of vitamin C supplementation to improve the quality of regenerative tissues for stem cell therapies.

Vitamin C Prevents and Reverses Stem Cell Aging

An important function of vitamin C treatment during reprogramming and in the maintenance of pluripotent stem cells is the prevention of senescence and aging in long-term *in vitro* cultures. Vitamin C treatment silences the *Ink4/Arf* locus in ESCs and iPSCs by increasing the activity of H3K36 demethylases [35,36], which has the potential to rejuvenate aged cells [4,98]. The regulation of telomere length and the expression of telomerase-related genes that regulate telomerase activity can also have a profound effect on cell aging. Vitamin C treatment of human iPSC- and ESC-derived cardiomyocytes (CMs) reverses aging phenotypes and maintains the multipotent potential of MSCs *in vitro* by increasing telomerase activity and the expression of genes encoding telomerase-related RNA and protein components that protect telomere stability [97,99].

In patients with Werner syndrome (WS), loss-of-function mutations of the *WRN* gene, which is involved in DNA repair and telomere maintenance, causes premature aging phenotypes such as gray hair, osteoporosis, diabetes, and cancer [100,101]. Using a human MSC model, the major cellular defects of WS can be observed, including decreased proliferation, accelerated senescence and growth arrest, telomere attrition, and loss of stem cell viability [101]. Treatment with vitamin C slows down telomere shortening in WS MSCs and causes downregulation of aging markers such as the senescence-inducing p16/Ink4a protein and reduces the expression of pro-inflammatory cytokines IL-6 and IL-8 that are associated with a senescence phenotype [101,102]. Global changes in the transcriptome of genes and pathways that suppress accelerated aging in WS MSCs upon vitamin C treatment could be a direct effect of enhanced activity of α-KGDD family proteins. TETs are known to play important roles in telomere maintenance, DNA repair, and genomic stability [103–106]. Mouse ESCs depleted of *Tet1* and/or *Tet2* exhibit shorter telomeres, chromosomal instability, and reduced telomere recombination [106], whereas *Tet1/2/3*-triple KO ESCs have heterogeneous telomere lengths, increased frequency of telomere loss, chromosome segregation defects, and increased telomere–sister chromatid exchange that may be a compensatory mechanism to counteract telomere shortening [103–106]. Subtelomeres are also hypermethylated in TET-depleted ESCs, which may impede telomere elongation by recombination [106]. Vitamin C–mediated regulation of telomerase activity to maintain pluripotency and reverse aging phenotypes in genetic disorders such as WS also suggests that vitamin C could be used to rejuvenate aging tissues and prevent premature aging of normal differentiated adult cell types.

VITAMIN C–MEDIATED EPIGENETIC REGULATION OF HEMATOPOIETIC STEM CELLS

It is well known that vitamin C is important for maintaining a healthy immune system [107], but the ability of vitamin C to modulate HSC function through the regulation of epigenetic factors has only recently gained attention. TET proteins are important regulators of DNA methylation fidelity in HSCs, and loss of TET function occurs in a variety of hematopoietic malignancies (reviewed in [21,108]). A role for vitamin C in regulating TET enzymatic activity to maintain physiologic levels of DNA methylation and hydroxymethylation in HSCs therefore has important implications for the prevention and treatment of hematopoietic malignancy (Figure 6.3).

TET Loss of Function Drives Hematopoietic Malignancy

Decreased expression of TET proteins and loss of 5 hmC have been reported in multiple solid tumors [109–111]; however, hematopoietic malignancies exhibit the highest frequency of loss of function mutations in *TET* genes and primarily in *TET2* [21,108,112]. *TET1* participates as a translocation partner in MLL-rearranged acute myeloid leukemia (AML) [113], whereas *TET2* deletions and mutations have been detected in up to 30% of myelodysplastic syndrome (MDS) [114–116], 50% of chronic myelomonocytic leukemia (CMML) [116], 10%–20% of *de novo* AML, and in patients who transform from MPD to AML [117,118]. *TET2* mutations are also detected at a significant frequency in patients with B- or T-cell lymphoma [119–124] and in 30% of individuals with clonal hematopoiesis (CH), a premalignant state seen in approximately 10% of the healthy elderly population that increases their risk of progression to AML [114,125].

TET loss of function has been modeled genetically in mice, confirming that impaired activity of these enzymes is associated with HSC transformation and malignancy. The three TET proteins are differentially expressed in the adult HSCs, and hematopoietic cell lineages. TET1 is most highly expressed in long-term HSCs and developing B cells, whereas TET2 and TET3 are more highly expressed in short-term stem, progenitor, and myeloid lineage cells [126–129].

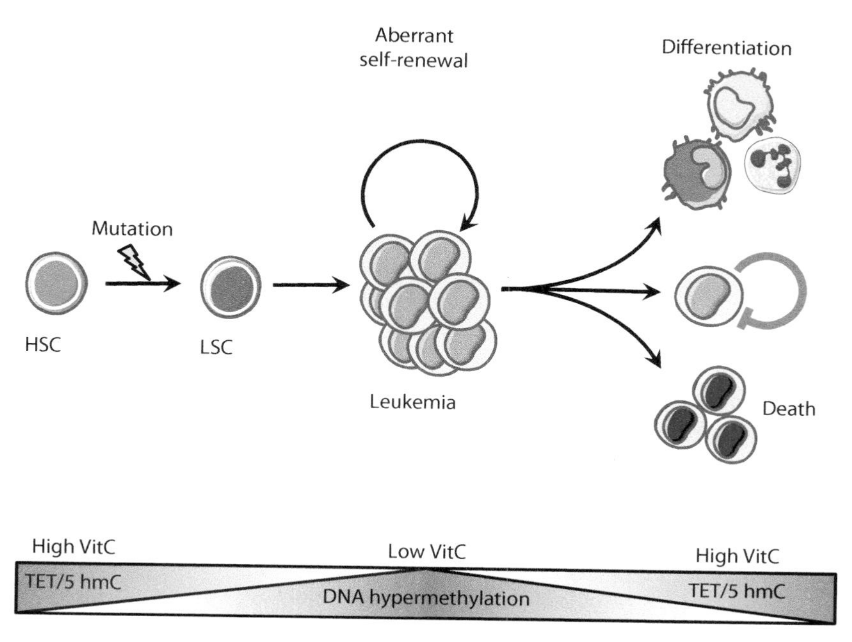

Figure 6.3. Vitamin C levels influence leukemia progression. Hematopoietic stem cells (HSCs) require higher levels of intracellular vitamin C compared to more differentiated cell types. Deficiency in vitamin C can impair TET function and promote the expansion of leukemia stem cells (LSCs) and leukemia. Low vitamin C levels or TET deficiency leads to decreased 5 hmC and DNA hypermethylation. High levels of vitamin C can increase TET enzymatic activity leading to a block in aberrant self-renewal and differentiation or death of leukemia cells.

These lineage-specific differences in TET expression may explain why *Tet1* deficiency in mice leads to aberrant self-renewal and expansion of HSCs with a B-cell lineage bias [126,127], whereas *Tet2* or combined *Tet2* and *Tet3* deficiency causes aberrant self-renewal with a myeloid lineage bias [129–131]. In addition, the combined loss of *Tet1/2* restricts malignancy to the B-cell lineage, whereas *Tet2/3* deficiency causes an accelerated AML [127,128]. In all of these mouse models, deficiency of TET proteins causes loss of 5 hmC in the genome of HSCs, DNA hypermethylation, and genomic instability.

Vitamin C Deficiency Alters Hematopoietic Stem Cell Function and Promotes Leukemia Progression

Intracellular vitamin C levels and expression of the sodium-dependent vitamin C transporter 2 (SVCT2) gene (*Slc23a2*) are remarkably higher in HSCs than restricted hematopoietic progenitors or more differentiated immune cell types, suggesting an increased requirement for vitamin C uptake in the stem cell compartment [22]. The effect of vitamin C deficiency in the regulation of HSC function has been modeled recently in mice that carry a spontaneous mutation in the gulonolactone (L-) oxidase (*Gulo*) locus and are used to model systemic vitamin C deficiency and dietary dependence [132]. Sodium L-ascorbate supplementation in the drinking water (3.3 g/L) is sufficient to maintain wild-type levels of vitamin C (80 μM) in the plasma of $Gulo^{-/-}$ mice; however, minimal supplementation (0.33 g/L) causes a rapid decrease in plasma concentration to subscurvy levels (less than 30 μM) [133,134]. Vitamin C restriction in the diet of $Gulo^{-/-}$ mice causes hematopoietic defects including reduced bone mass and smaller spleens and thymi compared to heterozygous or wild-type controls [135,136]. In addition, vitamin C deficiency in $Gulo^{-/-}$ mice leads to increased HSC frequency and a loss of 5 hmC in the genome that mimics the effect of TET deficiency in HSCs, an effect that could be reversed on dietary vitamin C administration [22]. Mice deficient in Svct2 have also been generated [136,137]. *Svct2* knockout in bone marrow cells cooperates with $Flt3^{ITD}$ oncogene overexpression to accelerate leukemia progression in mice [22]. Vitamin C deficiency also exacerbates 5 hmC loss in HSCs with heterozygous or homozygous deletion of *Tet2*, suggesting that systemic vitamin C depletion could impair the activity of all TET proteins in HSCs, including TET1 and TET3 [22]. Intriguingly, the majority of patients with hematologic malignancies (up to 58%) are vitamin C deficient compared to normal healthy controls [138,139], and chemotherapy or HSC transplantation can cause vitamin C levels to decrease even further in these patients [140]. If vitamin C deficiency can exacerbate leukemia progression in mouse models, a decreased amount of circulating vitamin C in patients may further impair TET function. Given that loss of function in *TET2* is known to confer a poor prognosis in AML [116,141], vitamin C deficiency could fuel increased disease aggressiveness or risk of relapse even in non-*TET* mutant patients.

Vitamin C Treatment Restores TET Function in Leukemia Stem Cells

TET2 mutations in patients are almost always heterozygous [112,116,142,143], suggesting that enhancing residual TET2 activity, encoded by the remaining wild-type *TET2* allele, or restoring the activity of functionally defective mutant TET2 proteins, could be a viable therapeutic strategy for the treatment of patients with *TET2* mutant AML.

Using genetic mouse models of reversible RNA interference, restoration of endogenous *Tet2* expression levels in *Tet2*-knockdown cells blocks aberrant HSC self-renewal, increases 5 hmC, promotes DNA hypomethylation, and upregulates a myeloid differentiation gene expression signature [23]. Given its effects in fibroblast reprogramming it was hypothesized that vitamin C treatment might enhance the enzymatic activity of residual TET2 or TET1/3, to promote 5 hmC formation and DNA demethylation in the MDS/AML genome at hypermethylated loci [23]. Vitamin C mimics genetic *Tet2* restoration, causing increased 5 hmC formation, a block in aberrant self-renewal in both human or mouse HSPCs, and suppression of disease progression in *Tet2*-deficient mice [23]. In HSPCs, *Tet2* and *Tet3* account for greater than 95% of *Tet* expression [23,128], and combined loss of *Tet2 and Tet3* in mouse HSPCs almost completely deplet es 5 hmC in the genome that cannot be increased upon vitamin C treatment [23]. These studies suggest that the effect of vitamin C treatment and 5 hmC formation is dependent on residual TET expression and enzymatic activity.

Vitamin C–Mediated Erasure of Epigenetic Memory in Leukemia Cells

DNA hypermethylation has been well documented to cause silencing at the loci of tumor suppressor genes (reviewed by [144]) and is a hallmark of patients with myeloid malignancies such as MDS and AML [117,145]. *TET2* mutations are associated with a DNA hypermethylation phenotype [117,146], an increased risk of MDS progression [147,148], and a poor prognosis in AML [141,149]. Studies of healthy aged individuals with clonal hematopoiesis (CH) have revealed that a disease-associated DNA methylation signature already exists in premalignant hematopoietic cells [150]. Epigenetic regulators such as *DNMT3a* and *TET2* are the most frequently mutated genes in normal individuals with CH [125,151]. For *TET2* mutant CH individuals, the ratio of 5 hmC to 5 mC has been shown to track with *TET2* variant allele frequency, with higher mutant VAF leading to lower 5 hmC [150].

Isocitrate dehydrogenases (IDH1/2) are enzymes that generate α-KG, making them important players in the regulation of α-KGDD activity and therefore TET function. *IDH* mutations are frequently found in patients with CH, leukemia, and lymphoma; are mutually exclusive with *TET2* mutations; and confer an overlapping DNA hypermethylation signature with *TET2* mutant leukemia [117,152,153]. *IDH* mutation leads to a gain of function, where instead of generating α-KG, mutant enzymes produce 2-hydroxyglutarate (2-HG), an oncometabolite that impairs the catalytic activity of α-KGDDs causing a deficiency in TET function [154–157]. Using *IDH* mutant mouse models, it was shown that compared to *IDH* wild-type cells, *IDH* mutant HSPCs lose 5 hmC and develop a DNA hypermethylation phenotype, aberrant self-renewal, and can cooperate with other oncogenic lesions to drive an aggressive AML [158–160]. Vitamin C treatment has been tested on *IDH1* mutant mouse leukemia cells, where it was shown to induce a Tet2-dependent gain of 5 hmC, loss of 5 mC, and the upregulation of gene expression signatures that correlate with decreased self-renewal and increased differentiation toward a mature myeloid phenotype [161].

Wilms tumor-1 (*WT1*) mutations are also a frequent occurrence in patients with AML and exhibit mutual exclusivity with *TET2* and *IDH* mutations [162]. *WT1* physically interacts with TET2 and TET3 [162] and recruits TET2 to its target genes [163].

WT1-mutant AMLs lose normal 5 hmC patterning and enrichment at *TET2*-regulated genomic regions such as enhancers, gene bodies, and distal regulatory elements, most likely because of impaired chromatin recruitment of *TET2* [162,163]. The epigenetic phenotypes of *IDH*, *WT1*, and *TET2* mutant AML suggest that a convergent mechanism of HSPC transformation and leukemia maintenance involves loss of function in *TET2*, leading to disordered DNA hydroxymethylation and DNA hypermethylation. The effect of vitamin C treatment in *WT1* leukemia has not yet been tested; however, the responsiveness of *IDH* mutant leukemia cells to vitamin C (despite being depleted of α-KG) suggests that enhancing the residual amount of functional wild-type *TET2* in *WT1* mutant AML could be sufficient to restore epigenetic differentiation cues.

DNA Hypomethylating Therapies and Vitamin C

In human leukemia cell lines, treatment with vitamin C causes widespread DNA hypomethylation [23,138], similar to what has been observed in the genome of ESCs [24]. Vitamin C treatment also induces the expression of a *TET2*-dependent gene expression signature in human leukemia cell lines [23]. Inhibitors of DNA methyltransferase (DNMTis), such as 5-azacytidine and decitabine, are approved by the U.S. Food and Drug Administration for the treatment of MDS that act by reversing aberrant DNA hypermethylation [164–166]. Vitamin C treatment synergizes with decitabine in a *TET2*-dependent manner to increase 5 hmC, drive DNA hypomethylation, and upregulate the expression of endogenous retroviral genes, triggering an innate viral mimicry response that promotes apoptosis of several human cancer cells lines including AML [138]. Restoring TET function by vitamin C administration may, in combination with DNMTi therapy, help to erase epigenetic memory of the leukemic cell state and reprogram the cells such that they can reacquire their normal differentiation potential and tumor suppressive gene expression programs.

Vitamin C treatment also catalyzes TET-mediated iterative oxidation of 5 hmC to generate 5 fC and 5 caC. These modified bases mimic oxidative DNA damage and trigger a base excision repair response that is dependent on PARP proteins and facilitates active DNA demethylation [68,69]. Vitamin C treatment in combination with the PARP inhibitor, Olaparib, enhances the killing of human AML cells greater than either agent alone [8]. TET-mediated DNA oxidation induced by vitamin C can therefore create a synthetic lethality, where AML cells being forced to undergo active DNA demethylation renders them hypersensitive to PARP inhibition that can be exploited as a novel combination therapy for the treatment of leukemia.

CONCLUDING REMARKS

The potential for vitamin C to modulate the epigenome as a natural, nontoxic, enhancer of somatic cell reprogramming and for the prevention and treatment of cancer has important implications for human biology. Vitamin C, by acting as a cofactor of epigenetic regulators such as JHDM and TET proteins, has become an essential media component for generating high-quality and efficiently reprogrammed somatic cells, where the erasure of epigenetic memory is a key step to restoring differentiation plasticity. The combination of epigenetic and nonepigenetic influences of vitamin C, through its ability to regulate diverse families of α-KGDDs, including collagen prolyl hydroxylases, also makes it an important adjuvant for stem cell therapies in regenerative medicine. Furthermore, given that loss of function in epigenetic regulators is a hallmark of cancer and a driver of disease progression, the ability of vitamin C to suppress leukemia progression by enhancing TET enzymatic activity could herald a new era of epigenetic therapy for the treatment of hematopoietic malignancies.

REFERENCES

1. Takahashi, K. and Yamanaka, S. 2006 Induction of pluripotent stem cells from mouse embryonic and adult fibroblast cultures by defined factors. *Cell* 126, 663–676.
2. Zhao, X. Y., Li, W., Lv, Z., Liu, L., Tong, M., Hai, T., Hao, J. et al. 2009 iPS cells produce viable mice through tetraploid complementation. *Nature* 461, 86–90.
3. Marion, R. M., Strati, K., Li, H., Murga, M., Blanco, R., Ortega, S., Fernandez-Capetillo, O., Serrano, M. and Blasco, M. A. 2009 A p53-mediated DNA damage response limits reprogramming to ensure iPS cell genomic integrity. *Nature* 460, 1149–1153.
4. Li, H., Collado, M., Villasante, A., Strati, K., Ortega, S., Canamero, M., Blasco, M. A. and

Serrano, M. 2009 The Ink4/Arf locus is a barrier for iPS cell reprogramming. *Nature* 460, 1136–1139.

5. Eminli, S., Foudi, A., Stadtfeld, M., Maherali, N., Ahfeldt, T., Mostoslavsky, G., Hock, H. and Hochedlinger, K. 2009 Differentiation stage determines potential of hematopoietic cells for reprogramming into induced pluripotent stem cells. *Nat. Genet.* 41, 968–976.
6. Maherali, N., Ahfeldt, T., Rigamonti, A., Utikal, J., Cowan, C. and Hochedlinger, K. 2008 A high-efficiency system for the generation and study of human induced pluripotent stem cells. *Cell Stem Cell* 3, 340–345.
7. Kim, K., Doi, A., Wen, B., Ng, K., Zhao, R., Cahan, P., Kim, J. et al. 2010 Epigenetic memory in induced pluripotent stem cells. *Nature* 467, 285–290.
8. Hussein, S. M., Batada, N. N., Vuoristo, S., Ching, R. W., Autio, R., Narva, E., Ng, S. et al. 2011 Copy number variation and selection during reprogramming to pluripotency. *Nature* 471, 58–62.
9. Gore, A., Li, Z., Fung, H. L., Young, J. E., Agarwal, S., Antosiewicz-Bourget, J., Canto, I. et al. 2011 Somatic coding mutations in human induced pluripotent stem cells. *Nature* 471, 63–67.
10. Stadtfeld, M., Apostolou, E., Akutsu, H., Fukuda, A., Follett, P., Natesan, S., Kono, T., Shioda, T. and Hochedlinger, K. 2010 Aberrant silencing of imprinted genes on chromosome 12qF1 in mouse induced pluripotent stem cells. *Nature* 465, 175–181.
11. Wu, S. M. and Hochedlinger, K. 2011 Harnessing the potential of induced pluripotent stem cells for regenerative medicine. *Nat. Cell Biol.* 13, 497–505.
12. Stadtfeld, M. and Hochedlinger, K. 2010 Induced pluripotency: History, mechanisms, and applications. *Genes Dev.* 24, 2239–2263.
13. Polo, J. M., Liu, S., Figueroa, M. E., Kulalert, W., Eminli, S., Tan, K. Y., Apostolou, E. et al. 2010 Cell type of origin influences the molecular and functional properties of mouse induced pluripotent stem cells. *Nat. Biotechnol.* 28, 848–855.
14. Stadtfeld, M., Apostolou, E., Ferrari, F., Choi, J., Walsh, R. M., Chen, T., Ooi, S. S. et al. (2012). Ascorbic acid prevents loss of Dlk1-Dio3 imprinting and facilitates generation of all-iPS cell mice from terminally differentiated B cells. *Nat. Genet.* 44, 398–405, S391–392.
15. Esteban, M. A., Wang, T., Qin, B., Yang, J., Qin, D., Cai, J., Li, W. et al. 2010 Vitamin C enhances the generation of mouse and human induced pluripotent stem cells. *Cell Stem Cell* 6, 71–79.
16. Wang, T., Chen, K., Zeng, X., Yang, J., Wu, Y., Shi, X., Qin, B. et al. 2011 The histone demethylases Jhdm1a/1b enhance somatic cell reprogramming in a vitamin-C-dependent manner. *Cell Stem Cell* 9, 575–587.
17. Young, J. I., Zuchner, S. and Wang, G. 2015 Regulation of the epigenome by vitamin C. *Annu. Rev. Nutr.* 35, 545–564.
18. Blaschke, K., Ebata, K. T., Karimi, M. M., Zepeda-Martinez, J. A., Goyal, P., Mahapatra, S., Tam, A. et al. 2013 Vitamin C induces Tet-dependent DNA demethylation and a blastocyst-like state in ES cells. *Nature* 500, 222–226.
19. Yin, R., Mao, S. Q., Zhao, B., Chong, Z., Yang, Y., Zhao, C., Zhang, D. et al. 2013 Ascorbic acid enhances Tet-mediated 5-methylcytosine oxidation and promotes DNA demethylation in mammals. *J. Am. Chem. Soc.* 135, 10396–10403.
20. Cimmino, L., Abdel-Wahab, O., Levine, R. L. and Aifantis, I. 2011 TET family proteins and their role in stem cell differentiation and transformation. *Cell Stem Cell* 9, 193–204.
21. Guillamot, M., Cimmino, L. and Aifantis, I. 2016 The impact of DNA methylation in hematopoietic malignancies. *Trends Cancer* 2, 70–83.
22. Agathocleous, M., Meacham, C. E., Burgess, R. J., Piskounova, E., Zhao, Z., Crane, G. M., Cowin, B. L. et al. 2017 Ascorbate regulates haematopoietic stem cell function and leukaemogenesis. *Nature* 549, 476–481.
23. Cimmino, L., Dolgalev, I., Wang, Y., Yoshimi, A., Martin, G. H., Wang, J., Ng, V. et al. (2017). Restoration of TET2 function blocks aberrant self-renewal and leukemia progression. *Cell* 170, 1079–1095, e1020.
24. Chung, T. L., Brena, R. M., Kolle, G., Grimmond, S. M., Berman, B. P., Laird, P. W., Pera, M. F. and Wolvetang, E. J. 2010 Vitamin C promotes widespread yet specific DNA demethylation of the epigenome in human embryonic stem cells. *Stem Cells* 28, 1848–1855.
25. Cloos, P. A., Christensen, J., Agger, K. and Helin, K. 2008 Erasing the methyl mark: Histone demethylases at the center of cellular differentiation and disease. *Genes Dev.* 22, 1115–1140.
26. Tsukada, Y., Fang, J., Erdjument-Bromage, H., Warren, M. E., Borchers, C. H., Tempst, P. and Zhang, Y. 2006 Histone demethylation by a

family of JmjC domain-containing proteins. *Nature* 439, 811–816.

27. Klose, R. J., Kallin, E. M. and Zhang, Y. 2006 JmjC-domain-containing proteins and histone demethylation. *Nat. Rev. Genet.* 7, 715–727.
28. Clifton, I. J., McDonough, M. A., Ehrismann, D., Kershaw, N. J., Granatino, N. and Schofield, C. J. 2006 Structural studies on 2-oxoglutarate oxygenases and related double-stranded beta-helix fold proteins. *J. Inorg. Biochem.* 100, 644–669.
29. Monfort, A. and Wutz, A. 2013 Breathing-in epigenetic change with vitamin C. *EMBO Rep.* 14, 337–346.
30. Pedersen, M. T. and Helin, K. 2010 Histone demethylases in development and disease. *Trends Cell Biol.* 20, 662–671.
31. Chen, J., Liu, H., Liu, J., Qi, J., Wei, B., Yang, J., Liang, H. et al. 2013 H3K9 methylation is a barrier during somatic cell reprogramming into iPSCs. *Nat. Genet.* 45, 34–42.
32. Ebata, K. T., Mesh, K., Liu, S., Bilenky, M., Fekete, A., Acker, M. G., Hirst, M., Garcia, B. A. and Ramalho-Santos, M. 2017 Vitamin C induces specific demethylation of H3K9me2 in mouse embryonic stem cells via Kdm3a/b. *Epigenetics Chromatin* 10, 36.
33. Mansour, A. A., Gafni, O., Weinberger, L., Zviran, A., Ayyash, M., Rais, Y., Krupalnik, V. et al. 2012 The H3K27 demethylase Utx regulates somatic and germ cell epigenetic reprogramming. *Nature* 488, 409–413.
34. Hanna, J., Saha, K., Pando, B., van Zon, J., Lengner, C. J., Creyghton, M. P., van Oudenaarden, A. and Jaenisch, R. 2009 Direct cell reprogramming is a stochastic process amenable to acceleration. *Nature* 462, 595–601.
35. He, J., Kallin, E. M., Tsukada, Y. and Zhang, Y. 2008 The H3K36 demethylase Jhdm1b/Kdm2b regulates cell proliferation and senescence through p15(Ink4b). *Nat. Struct. Mol. Biol.* 15, 1169–1175.
36. Tzatsos, A., Pfau, R., Kampranis, S. C. and Tsichlis, P. N. 2009 Ndy1/KDM2B immortalizes mouse embryonic fibroblasts by repressing the Ink4a/Arf locus. *Proc. Natl. Acad. Sci. USA* 106, 2641–2646.
37. Tahiliani, M., Koh, K. P., Shen, Y., Pastor, W. A., Bandukwala, H., Brudno, Y., Agarwal, S. et al. 2009 Conversion of 5-methylcytosine to 5-hydroxymethylcytosine in mammalian DNA by MLL partner TET1. *Science* 324, 930–935.
38. Ito, S., D'Alessio, A. C., Taranova, O. V., Hong, K., Sowers, L. C. and Zhang, Y. 2010 Role of Tet proteins in 5 mC to 5 hmC conversion, ES-cell self-renewal and inner cell mass specification. *Nature* 466, 1129–1133.
39. Zhang, H., Zhang, X., Clark, E., Mulcahey, M., Huang, S. and Shi, Y. G. 2010 TET1 is a DNA-binding protein that modulates DNA methylation and gene transcription via hydroxylation of 5-methylcytosine. *Cell Res.* 20, 1390–1393.
40. Koh, K. P., Yabuuchi, A., Rao, S., Huang, Y., Cunniff, K., Nardone, J., Laiho, A. et al. 2011 Tet1 and Tet2 regulate 5-hydroxymethylcytosine production and cell lineage specification in mouse embryonic stem cells. *Cell Stem Cell* 8, 200–213.
41. Dawlaty, M. M., Breiling, A., Le, T., Raddatz, G., Barrasa, M. I., Cheng, A. W., Gao, Q. et al. 2013 Combined deficiency of Tet1 and Tet2 causes epigenetic abnormalities but is compatible with postnatal development. *Dev. Cell* 24, 310–323.
42. Doege, C. A., Inoue, K., Yamashita, T., Rhee, D. B., Travis, S., Fujita, R., Guarnieri, P. et al. 2012 Early-stage epigenetic modification during somatic cell reprogramming by Parp1 and Tet2. *Nature* 488, 652–655.
43. Costa, Y., Ding, J., Theunissen, T. W., Faiola, F., Hore, T. A., Shliaha, P. V., Fidalgo, M. et al. 2013 NANOG-dependent function of TET1 and TET2 in establishment of pluripotency. *Nature* 495, 370–374.
44. Gao, Y., Chen, J., Li, K., Wu, T., Huang, B., Liu, W., Kou, X. et al. 2013 Replacement of Oct4 by Tet1 during iPSC induction reveals an important role of DNA methylation and hydroxymethylation in reprogramming. *Cell Stem Cell* 12, 453–469.
45. Chen, Q., Chen, Y., Bian, C., Fujiki, R. and Yu, X. 2013 TET2 promotes histone O-GlcNAcylation during gene transcription. *Nature* 493, 561–564.
46. Minor, E. A., Court, B. L., Young, J. I. and Wang, G. 2013 Ascorbate induces ten-eleven translocation (Tet) methylcytosine dioxygenase-mediated generation of 5-hydroxymethylcytosine. *J. Biol. Chem.* 288, 13669–13674.
47. Wu, H. and Zhang, Y. 2011 Tet1 and 5-hydroxymethylation: A genome-wide view in mouse embryonic stem cells. *Cell Cycle* 10.
48. Williams, K., Christensen, J., Pedersen, M. T., Johansen, J. V., Cloos, P. A., Rappsilber, J. and Helin, K. 2011 TET1 and hydroxymethylcytosine in transcription and DNA methylation fidelity. *Nature* 473, 343–348.

49. Xu, Y., Wu, F., Tan, L., Kong, L., Xiong, L., Deng, J., Barbera, A. J. et al. 2011 Genome-wide regulation of 5 hmC, 5 mC, and gene expression by Tet1 hydroxylase in mouse embryonic stem cells. *Mol. Cell* 42, 451–464.
50. Deplus, R., Delatte, B., Schwinn, M. K., Defrance, M., Mendez, J., Murphy, N., Dawson, M. A. et al. 2013 TET2 and TET3 regulate GlcNAcylation and H3K4 methylation through OGT and SET1/COMPASS. *EMBO J.* 32, 645–655.
51. Vella, P., Scelfo, A., Jammula, S., Chiacchiera, F., Williams, K., Cuomo, A., Roberto, A. et al. 2013 Tet proteins connect the O-linked N-acetylglucosamine transferase Ogt to chromatin in embryonic stem cells. *Mol. Cell* 49, 645–656.
52. Dawlaty, M. M., Breiling, A., Le, T., Barrasa, M. I., Raddatz, G., Gao, Q., Powell, B. E. et al. 2014 Loss of Tet enzymes compromises proper differentiation of embryonic stem cells. *Dev. Cell* 29, 102–111.
53. Schwarz, B. A., Bar-Nur, O., Silva, J. C. and Hochedlinger, K. 2014 Nanog is dispensable for the generation of induced pluripotent stem cells. *Curr. Biol.* 24, 347–350.
54. Hore, T. A., von Meyenn, F., Ravichandran, M., Bachman, M., Ficz, G., Oxley, D., Santos, F., Balasubramanian, S., Jurkowski, T. P. and Reik, W. 2016 Retinol and ascorbate drive erasure of epigenetic memory and enhance reprogramming to naive pluripotency by complementary mechanisms. *Proc. Natl. Acad. Sci. USA* 113, 12202–12207.
55. Chen, J., Guo, L., Zhang, L., Wu, H., Yang, J., Liu, H., Wang, X. et al. 2013 Vitamin C modulates TET1 function during somatic cell reprogramming. *Nat. Genet.* 45, 1504–1509.
56. Keith, B. and Simon, M. C. 2007 Hypoxia-inducible factors, stem cells, and cancer. *Cell* 129, 465–472.
57. Mohyeldin, A., Garzon-Muvdi, T. and Quinones-Hinojosa, A. 2010 Oxygen in stem cell biology: A critical component of the stem cell niche. *Cell Stem Cell* 7, 150–161.
58. Semenza, G. L. 1999 Perspectives on oxygen sensing. *Cell.* 98, 281–284.
59. Yoshida, Y., Takahashi, K., Okita, K., Ichisaka, T. and Yamanaka, S. 2009 Hypoxia enhances the generation of induced pluripotent stem cells. *Cell Stem Cell* 5, 237–241.
60. Panopoulos, A. D., Yanes, O., Ruiz, S., Kida, Y. S., Diep, D., Tautenhahn, R., Herrerias, A. et al. 2012 The metabolome of induced pluripotent stem cells reveals metabolic changes occurring in somatic cell reprogramming. *Cell Res.* 22, 168–177.
61. Folmes, C. D., Nelson, T. J., Martinez-Fernandez, A., Arrell, D. K., Lindor, J. Z., Dzeja, P. P., Ikeda, Y., Perez-Terzic, C. and Terzic, A. 2011 Somatic oxidative bioenergetics transitions into pluripotency-dependent glycolysis to facilitate nuclear reprogramming. *Cell Metab.* 14, 264–271.
62. Zhou, W., Choi, M., Margineantu, D., Margaretha, L., Hesson, J., Cavanaugh, C., Blau, C. A. et al. 2012 HIF1alpha induced switch from bivalent to exclusively glycolytic metabolism during ESC-to-EpiSC/hESC transition. *EMBO J.* 31, 2103–2116.
63. Mathieu, J., Zhou, W., Xing, Y., Sperber, H., Ferreccio, A., Agoston, Z., Kuppusamy, K. T., Moon, R. T. and Ruohola-Baker, H. 2014 Hypoxia-inducible factors have distinct and stage-specific roles during reprogramming of human cells to pluripotency. *Cell Stem Cell* 14, 592–605.
64. Lando, D., Peet, D. J., Gorman, J. J., Whelan, D. A., Whitelaw, M. L. and Bruick, R. K. 2002 FIH-1 is an asparaginyl hydroxylase enzyme that regulates the transcriptional activity of hypoxia-inducible factor. *Genes Dev.* 16, 1466–1471.
65. Kawada, H., Kaneko, M., Sawanobori, M., Uno, T., Matsuzawa, H., Nakamura, Y., Matsushita, H. and Ando, K. 2013 High concentrations of L-ascorbic acid specifically inhibit the growth of human leukemic cells via downregulation of HIF-1alpha transcription. *PloS One* 8, e62717.
66. Nordstrand, L. M., Svard, J., Larsen, E., Nilsen, A., Ougland, R., Furu, K., Lien, G. F. et al. 2010 Mice lacking Alkbh1 display sex-ratio distortion and unilateral eye defects. *PloS One* 5, e13827.
67. Ougland, R., Lando, D., Jonson, I., Dahl, J. A., Moen, M. N., Nordstrand, L. M., Rognes, T. et al. 2012 ALKBH1 is a histone H2A dioxygenase involved in neural differentiation. *Stem Cells* 30, 2672–2682.
68. Guenther, M. G., Frampton, G. M., Soldner, F., Hockemeyer, D., Mitalipova, M., Jaenisch, R. and Young, R. A. 2010 Chromatin structure and gene expression programs of human embryonic and induced pluripotent stem cells. *Cell Stem Cell* 7, 249–257.
69. Ma, Y., Li, C., Gu, J., Tang, F., Li, C., Li, P., Ping, P., Yang, S., Li, Z. and Jin, Y. 2012 Aberrant gene expression profiles in pluripotent stem cells induced from fibroblasts of a Klinefelter syndrome patient. *J. Biol. Chem.* 287, 38970–38979.

70. Jia, G., Fu, Y., Zhao, X., Dai, Q., Zheng, G., Yang, Y., Yi, C. et al. 2011 N6-methyladenosine in nuclear RNA is a major substrate of the obesity-associated FTO. *Nat. Chem. Biol.* 7, 885–887.
71. Zheng, G., Dahl, J. A., Niu, Y., Fedorcsak, P., Huang, C. M., Li, C. J., Vagbo, C. B. et al. 2013 ALKBH5 is a mammalian RNA demethylase that impacts RNA metabolism and mouse fertility. *Mol. Cell* 49, 18–29.
72. Chen, T., Hao, Y. J., Zhang, Y., Li, M. M., Wang, M., Han, W., Wu, Y. et al. 2015 m(6)A RNA methylation is regulated by microRNAs and promotes reprogramming to pluripotency. *Cell Stem Cell* 16, 289–301.
73. Yu, X. X., Liu, Y. H., Liu, X. M., Wang, P. C., Liu, S., Miao, J. K., Du, Z. Q. and Yang, C. X. 2018 Ascorbic acid induces global epigenetic reprogramming to promote meiotic maturation and developmental competence of porcine oocytes. *Sci. Rep.* 8, 6132.
74. Scudellari, M. 2016 How iPS cells changed the world. *Nature* 534, 310–312.
75. Mandai, M., Watanabe, A., Kurimoto, Y., Hirami, Y., Morinaga, C., Daimon, T., Fujihara, M. et al. 2017 Autologous induced stem-cell-derived retinal cells for macular degeneration. *N. Engl. J. Med.* 376, 1038–1046.
76. Murry, C. E. and Keller, G. 2008 Differentiation of embryonic stem cells to clinically relevant populations: Lessons from embryonic development. *Cell* 132, 661–680.
77. D'Aniello, C., Cermola, F., Patriarca, E. J. and Minchiotti, G. 2017 Vitamin C in stem cell biology: Impact on extracellular matrix homeostasis and epigenetics. *Stem Cells Int.* 2017, 8936156.
78. Savitt, J. M., Dawson, V. L. and Dawson, T. M. 2006 Diagnosis and treatment of Parkinson disease: Molecules to medicine. *J. Clin. Invest.* 116, 1744–1754.
79. Wernig, M., Zhao, J. P., Pruszak, J., Hedlund, E., Fu, D., Soldner, F., Broccoli, V., Constantine-Paton, M., Isacson, O. and Jaenisch, R. 2008 Neurons derived from reprogrammed fibroblasts functionally integrate into the fetal brain and improve symptoms of rats with Parkinson's disease. *Proc. Natl. Acad. Sci. USA* 105, 5856–5861.
80. Wulansari, N., Kim, E. H., Sulistio, Y. A., Rhee, Y. H., Song, J. J. and Lee, S. H. 2017 Vitamin C-induced epigenetic modifications in donor NSCs establish midbrain marker expressions critical for cell-based therapy in Parkinson's disease. *Stem Cell Rep.* 9, 1192–1206.
81. He, X. B., Kim, M., Kim, S. Y., Yi, S. H., Rhee, Y. H., Kim, T., Lee, E. H. et al. 2015 Vitamin C facilitates dopamine neuron differentiation in fetal midbrain through TET1- and JMJD3-dependent epigenetic control manner. *Stem Cells* 33, 1320–1332.
82. Yorukoglu, A. C., Kiter, A. E., Akkaya, S., Satiroglu-Tufan, N. L. and Tufan, A. C. 2017 A concise review on the use of mesenchymal stem cells in cell sheet-based tissue engineering with special emphasis on bone tissue regeneration. *Stem Cells Int.* 2017, 2374161.
83. Hynes, K., Menicanin, D., Mrozik, K., Gronthos, S. and Bartold, P. M. 2014 Generation of functional mesenchymal stem cells from different induced pluripotent stem cell lines. *Stem Cells Dev.* 23, 1084–1096.
84. Moslem, M., Eberle, I., Weber, I., Henschler, R. and Cantz, T. 2015 Mesenchymal stem/stromal cells derived from induced pluripotent stem cells support CD34(pos) hematopoietic stem cell propagation and suppress inflammatory reaction. *Stem Cells Int.* 2015, 843058.
85. Soontararak, S., Chow, L., Johnson, V., Coy, J., Wheat, W., Regan, D. and Dow, S. 2018 Mesenchymal stem cells (MSC) derived from induced pluripotent stem cells (iPSC) equivalent to adipose-derived MSC in promoting intestinal healing and microbiome normalization in mouse inflammatory bowel disease model. *Stem Cells Transl. Med.* 7, 456–467.
86. Knott, L. and Bailey, A. J. 1998 Collagen cross-links in mineralizing tissues: A review of their chemistry, function, and clinical relevance. *Bone* 22, 181–187.
87. Yamauchi, M. and Sricholpech, M. 2012 Lysine post-translational modifications of collagen. *Essays Biochem.* 52, 113–133.
88. Juva, K., Prockop, D. J., Cooper, G. W. and Lash, J. W. 1966 Hydroxylation of proline and the intracellular accumulation of a polypeptide precursor of collagen. *Science* 152, 92–94.
89. Walmsley, A. R., Batten, M. R., Lad, U. and Bulleid, N. J. 1999 Intracellular retention of procollagen within the endoplasmic reticulum is mediated by prolyl 4-hydroxylase. *J. Biol. Chem.* 274, 14884–14892.
90. Kim, B., Choi, K. M., Yim, H. S. and Lee, M. G. 2013 Ascorbic acid enhances adipogenesis of 3T3-L1 murine preadipocyte through differential expression of collagens. *Lipids Health Dis.* 12, 182.
91. Geesin, J. C., Hendricks, L. J., Falkenstein, P. A., Gordon, J. S. and Berg, R. A. 1991

Regulation of collagen synthesis by ascorbic acid: Characterization of the role of ascorbate-stimulated lipid peroxidation. *Arch. Biochem. Biophys.* 290, 127–132.

92. Qiao, H., Bell, J., Juliao, S., Li, L. and May, J. M. 2009 Ascorbic acid uptake and regulation of type I collagen synthesis in cultured vascular smooth muscle cells. *J. Vasc. Res.* 46, 15–24.
93. Tajima, S. and Pinnell, S. R. 1982 Regulation of collagen synthesis by ascorbic acid. Ascorbic acid increases type I procollagen mRNA. *Biochem. Biophys. Res. Commun.* 106, 632–637.
94. Stephens, L. E., Sutherland, A. E., Klimanskaya, I. V., Andrieux, A., Meneses, J., Pedersen, R. A. and Damsky, C. H. 1995 Deletion of beta 1 integrins in mice results in inner cell mass failure and peri-implantation lethality. *Genes Dev.* 9, 1883–1895.
95. Li, S., Harrison, D., Carbonetto, S., Fassler, R., Smyth, N., Edgar, D. and Yurchenco, P. D. 2002 Matrix assembly, regulation, and survival functions of laminin and its receptors in embryonic stem cell differentiation. *J. Cell Biol.* 157, 1279–1290.
96. Hayashi, Y., Furue, M. K., Okamoto, T., Ohnuma, K., Myoishi, Y., Fukuhara, Y., Abe, T., Sato, J. D., Hata, R. and Asashima, M. 2007 Integrins regulate mouse embryonic stem cell self-renewal. *Stem Cells* 25, 3005–3015.
97. Wei, F., Qu, C., Song, T., Ding, G., Fan, Z., Liu, D., Liu, Y., Zhang, C., Shi, S. and Wang, S. 2012 Vitamin C treatment promotes mesenchymal stem cell sheet formation and tissue regeneration by elevating telomerase activity. *J. Cell Physiol.* 227, 3216–3224.
98. Krishnamurthy, J., Torrice, C., Ramsey, M. R., Kovalev, G. I., Al-Regaiey, K., Su, L. and Sharpless, N. E. 2004 Ink4a/Arf expression is a biomarker of aging. *J. Clin. Invest.* 114, 1299–1307.
99. Kim, Y. Y., Ku, S. Y., Huh, Y., Liu, H. C., Kim, S. H., Choi, Y. M. and Moon, S. Y. 2013 Anti-aging effects of vitamin C on human pluripotent stem cell-derived cardiomyocytes. *Age (Dordr)* 35, 1545–1557.
100. Burtner, C. R. and Kennedy, B. K. 2010 Progeria syndromes and ageing: What is the connection? *Nat. Rev. Mol. Cell Biol.* 11, 567–578.
101. Zhang, W., Li, J., Suzuki, K., Qu, J., Wang, P., Zhou, J., Liu, X. et al. 2015 Aging stem cells. A Werner syndrome stem cell model unveils heterochromatin alterations as a driver of human aging. *Science* 348, 1160–1163.
102. Zhang, W., Qu, J., Suzuki, K., Liu, G. H. and Izpisua Belmonte, J. C. 2013 Concealing cellular defects in pluripotent stem cells. *Trends Cell Biol.* 23, 587–592.
103. Kafer, G. R., Li, X., Horii, T., Suetake, I., Tajima, S., Hatada, I. and Carlton, P. M. 2016 5-hydroxymethylcytosine marks sites of DNA damage and promotes genome stability. *Cell Rep.* 14, 1283–1292.
104. Lazzerini-Denchi, E. and Sfeir, A. 2016 Stop pulling my strings—What telomeres taught us about the DNA damage response. Nature reviews. *Mol. Cell Biol.* 17, 364–378.
105. Lu, F., Liu, Y., Jiang, L., Yamaguchi, S. and Zhang, Y. 2014 Role of Tet proteins in enhancer activity and telomere elongation. *Genes Dev.* 28, 2103–2119.
106. Yang, J., Guo, R., Wang, H., Ye, X., Zhou, Z., Dan, J., Wang, H. et al. 2016 Tet enzymes regulate telomere maintenance and chromosomal stability of mouse ESCs. *Cell Rep.* 15, 1809–1821.
107. Carr, A. C. and Maggini, S. 2017 Vitamin C and immune function. *Nutrients* 9.
108. Bowman, R. L., Busque, L. and Levine, R. L. 2018 Clonal hematopoiesis and evolution to hematopoietic malignancies. *Cell Stem Cell* 22, 157–170.
109. Yang, H., Liu, Y., Bai, F., Zhang, J. Y., Ma, S. H., Liu, J., Xu, Z. D. et al. 2013 Tumor development is associated with decrease of TET gene expression and 5-methylcytosine hydroxylation. *Oncogene* 32, 663–669.
110. Haffner, M. C., Chaux, A., Meeker, A. K., Esopi, D. M., Gerber, J., Pellakuru, L. G., Toubaji, A. et al. 2011 Global 5-hydroxymethylcytosine content is significantly reduced in tissue stem/progenitor cell compartments and in human cancers. *Oncotarget* 2, 627–637.
111. Kudo, Y., Tateishi, K., Yamamoto, K., Yamamoto, S., Asaoka, Y., Ijichi, H., Nagae, G., Yoshida, H., Aburatani, H. and Koike, K. 2012 Loss of 5-hydroxymethylcytosine is accompanied with malignant cellular transformation. *Cancer Sci.* 103, 670–676.
112. Abdel-Wahab, O., Mullally, A., Hedvat, C., Garcia-Manero, G., Patel, J., Wadleigh, M., Malinge, S. et al. 2009 Genetic characterization of TET1, TET2, and TET3 alterations in myeloid malignancies. *Blood* 114, 144–147.
113. Lorsbach, R. B., Moore, J., Mathew, S., Raimondi, S. C., Mukatira, S. T. and Downing, J. R. 2003 TET1, a member of a novel protein family,

is fused to MLL in acute myeloid leukemia containing the t(10;11).(q22;q23). *Leukemia* 17, 637–641.

114. Delhommeau, F., Dupont, S., Della Valle, V., James, C., Trannoy, S., Masse, A., Kosmider, O. et al. 2009 Mutation in TET2 in myeloid cancers. *N. Engl. J. Med.* 360, 2289–2301.
115. Langemeijer, S. M., Kuiper, R. P., Berends, M., Knops, R., Aslanyan, M. G., Massop, M., Stevens-Linders, E. et al. 2009 Acquired mutations in TET2 are common in myelodysplastic syndromes. *Nat. Genet.* 41, 838–842.
116. Kosmider, O., Gelsi-Boyer, V., Ciudad, M., Racoeur, C., Jooste, V., Vey, N., Quesnel, B. et al. 2009 TET2 gene mutation is a frequent and adverse event in chronic myelomonocytic leukemia. *Haematologica* 94, 1676–1681.
117. Figueroa, M. E., Abdel-Wahab, O., Lu, C., Ward, P. S., Patel, J., Shih, A., Li, Y. et al. 2010 Leukemic IDH1 and IDH2 mutations result in a hypermethylation phenotype, disrupt TET2 function, and impair hematopoietic differentiation. *Cancer Cell* 18, 553–567.
118. Metzeler, K. H., Maharry, K., Radmacher, M. D., Mrozek, K., Margeson, D., Becker, H., Curfman, J. et al. 2011 TET2 mutations improve the new European LeukemiaNet risk classification of acute myeloid leukemia: A Cancer and Leukemia Group B study. *J. Clin. Oncol.: Off. J. Am. Soc. Clin. Oncol.* 29, 1373–1381.
119. Asmar, F., Punj, V., Christensen, J., Pedersen, M. T., Pedersen, A., Nielsen, A. B., Hother, C. et al. 2013 Genome-wide profiling identifies a DNA methylation signature that associates with TET2 mutations in diffuse large B-cell lymphoma. *Haematologica* 98, 1912–1920.
120. Quivoron, C., Couronne, L., Della Valle, V., Lopez, C. K., Plo, I., Wagner-Ballon, O., Do Cruzeiro, M. et al. 2011 TET2 inactivation results in pleiotropic hematopoietic abnormalities in mouse and is a recurrent event during human lymphomagenesis. *Cancer Cell* 20, 25–38.
121. Palomero, T., Couronne, L., Khiabanian, H., Kim, M. Y., Ambesi-Impiombato, A., Perez-Garcia, A., Carpenter, Z. et al. 2014 Recurrent mutations in epigenetic regulators, RHOA and FYN kinase in peripheral T cell lymphomas. *Nat Genet* 46, 166–170.
122. Sakata-Yanagimoto, M., Enami, T., Yoshida, K., Shiraishi, Y., Ishii, R., Miyake, Y., Muto, H. et al. 2014 Somatic RHOA mutation in angioimmunoblastic T cell lymphoma. *Nat. Genet.* 46, 171–175.
123. Odejide, O., Weigert, O., Lane, A. A., Toscano, D., Lunning, M. A., Kopp, N., Kim, S. et al. 2014 A targeted mutational landscape of angioimmunoblastic T-cell lymphoma. *Blood* 123, 1293–1296.
124. Couronne, L., Bastard, C. and Bernard, O. A. 2012 TET2 and DNMT3A mutations in human T-cell lymphoma. *N. Engl. J. Med.* 366, 95–96.
125. Jaiswal, S., Fontanillas, P., Flannick, J., Manning, A., Grauman, P. V., Mar, B. G., Lindsley, R. C. et al. 2014 Age-related clonal hematopoiesis associated with adverse outcomes. *N. Engl. J. Med.* 371, 2488–2498.
126. Cimmino, L., Dawlaty, M. M., Ndiaye-Lobry, D., Yap, Y. S., Bakogianni, S., Yu, Y., Bhattacharyya, S. et al. 2015 TET1 is a tumor suppressor of hematopoietic malignancy. *Nat. Immunol.* 16, 653–662.
127. Zhao, Z., Chen, L., Dawlaty, M. M., Pan, F., Weeks, O., Zhou, Y., Cao, Z. et al. 2015 Combined Loss of Tet1 and Tet2 Promotes B Cell, but Not Myeloid Malignancies, in Mice. *Cell Rep.* 13, 1692–1704.
128. An, J., Gonzalez-Avalos, E., Chawla, A., Jeong, M., Lopez-Moyado, I. F., Li, W., Goodell, M. A., Chavez, L., Ko, M. and Rao, A. 2015 Acute loss of TET function results in aggressive myeloid cancer in mice. *Nat. Commun.* 6, 10071.
129. Moran-Crusio, K., Reavie, L., Shih, A., Abdel-Wahab, O., Ndiaye-Lobry, D., Lobry, C., Figueroa, M. E. et al. 2011 Tet2 loss leads to increased hematopoietic stem cell self-renewal and myeloid transformation. *Cancer Cell* 20, 11–24.
130. Ko, M., Bandukwala, H. S., An, J., Lamperti, E. D., Thompson, E. C., Hastie, R., Tsangaratou, A., Rajewsky, K., Koralov, S. B. and Rao, A. 2011 Ten-Eleven-Translocation 2 (TET2) negatively regulates homeostasis and differentiation of hematopoietic stem cells in mice. *Proc. Natl. Acad. Sci. USA* 108, 14566–14571.
131. Li, Z., Cai, X., Cai, C., Wang, J., Zhang, W., Petersen, B. E., Yang, F. C. and Xu, M. 2011 Deletion of Tet2 in mice leads to dysregulated hematopoietic stem cells and subsequent development of myeloid malignancies. *Blood* 118, 4509–4518.
132. Maeda, N., Hagihara, H., Nakata, Y., Hiller, S., Wilder, J. and Reddick, R. 2000 Aortic wall damage in mice unable to synthesize ascorbic acid. *Proc. Natl. Acad. Sci. USA* 97, 841–846.
133. Kim, H., Bae, S., Yu, Y., Kim, Y., Kim, H. R., Hwang, Y. I., Kang, J. S. and Lee, W. J. 2012 The analysis of vitamin C concentration in organs of

gulo(−/−) mice upon vitamin C withdrawal. *Immune. Netw.* 12, 18–26.
134. Schleicher, R. L., Carroll, M. D., Ford, E. S. and Lacher, D. A. 2009 Serum vitamin C and the prevalence of vitamin C deficiency in the United States: 2003–2004 National Health and Nutrition Examination Survey (NHANES). *Am. J. Clin. Nutr.* 90, 1252–1263.
135. Beamer, W. G., Rosen, C. J., Bronson, R. T., Gu, W., Donahue, L. R., Baylink, D. J., Richardson, C. C., Crawford, G. C. and Barker, J. E. 2000 Spontaneous fracture (sfx): A mouse genetic model of defective peripubertal bone formation. *Bone* 27, 619–626.
136. Sotiriou, S., Gispert, S., Cheng, J., Wang, Y., Chen, A., Hoogstraten-Miller, S., Miller, G. F. et al. 2002 Ascorbic-acid transporter Slc23a1 is essential for vitamin C transport into the brain and for perinatal survival. *Nat. Med.* 8, 514–517.
137. Corpe, C. P., Tu, H., Eck, P., Wang, J., Faulhaber-Walter, R., Schnermann, J., Margolis, S. et al. 2010 Vitamin C transporter Slc23a1 links renal reabsorption, vitamin C tissue accumulation, and perinatal survival in mice. *J. Clin. Invest.* 120, 1069–1083.
138. Liu, M., Ohtani, H., Zhou, W., Orskov, A. D., Charlet, J., Zhang, Y. W., Shen, H. et al. 2016 Vitamin C increases viral mimicry induced by 5-aza-2′-deoxycytidine. *Proc. Natl. Acad. Sci. USA* 113, 10238–10244.
139. Huijskens, M. J., Wodzig, W. K., Walczak, M., Germeraad, W. T. and Bos, G. M. 2016 Ascorbic acid serum levels are reduced in patients with hematological malignancies. *Results Immunol.* 6, 8–10.
140. Nannya, Y., Shinohara, A., Ichikawa, M. and Kurokawa, M. 2014 Serial profile of vitamins and trace elements during the acute phase of allogeneic stem cell transplantation. *Biol. Blood Marrow Transplant* 20, 430–434.
141. Chou, W. C., Chou, S. C., Liu, C. Y., Chen, C. Y., Hou, H. A., Kuo, Y. Y., Lee, M. C. et al. 2011 TET2 mutation is an unfavorable prognostic factor in acute myeloid leukemia patients with intermediate-risk cytogenetics. *Blood* 118, 3803–3810.
142. Delhommeau, F., Dupont, S., Della Valle, V., James, C., Trannoy, S., Masse, A., Kosmider, O. et al. 2009 Mutation in TET2 in myeloid cancers. *N. Engl. J. Med.* 360, 2289–2301.
143. Langemeijer, S. M., Kuiper, R. P., Berends, M., Knops, R., Aslanyan, M. G., Massop, M., Stevens-Linders, E. et al. 2009 Acquired mutations in TET2 are common in myelodysplastic syndromes. *Nat. Genet.* 41, 838–842.
144. Baylin, S. B. and Jones, P. A. 2011 A decade of exploring the cancer epigenome—biological and translational implications. Nature reviews. *Cancer* 11, 726–734.
145. Figueroa, M. E., Lugthart, S., Li, Y., Erpelinck-Verschueren, C., Deng, X., Christos, P. J., Schifano, E. et al. 2010 DNA methylation signatures identify biologically distinct subtypes in acute myeloid leukemia. *Cancer Cell* 17, 13–27.
146. Issa, J. P. 2010 Epigenetic changes in the myelodysplastic syndrome. *Hematol. Oncol. Clin. North Am.* 24, 317–330.
147. Scharenberg, C., Giai, V., Pellagatti, A., Saft, L., Dimitriou, M., Jansson, M., Jadersten, M. et al. 2017 Progression in patients with low- and intermediate-1-risk del(5q) myelodysplastic syndromes is predicted by a limited subset of mutations. *Haematologica* 102, 498–508.
148. Jiang, Y., Dunbar, A., Gondek, L. P., Mohan, S., Rataul, M., O'Keefe, C., Sekeres, M., Saunthararajah, Y. and Maciejewski, J. P. 2009 Aberrant DNA methylation is a dominant mechanism in MDS progression to AML. *Blood* 113, 1315–1325.
149. Kosmider, O., Gelsi-Boyer, V., Cheok, M., Grabar, S., Della-Valle, V., Picard, F., Viguie, F. et al. 2009 TET2 mutation is an independent favorable prognostic factor in myelodysplastic syndromes (MDSs). *Blood* 114, 3285–3291.
150. Buscarlet, M., Provost, S., Zada, Y. F., Barhdadi, A., Bourgoin, V., Lepine, G., Mollica, L., Szuber, N., Dube, M. P. and Busque, L. 2017 DNMT3A and TET2 dominate clonal hematopoiesis and demonstrate benign phenotypes and different genetic predispositions. *Blood* 130, 753–762.
151. Genovese, G., Kahler, A. K., Handsaker, R. E., Lindberg, J., Rose, S. A., Bakhoum, S. F., Chambert, K. et al. 2014 Clonal hematopoiesis and blood-cancer risk inferred from blood DNA sequence. *N. Engl. J. Med.* 371, 2477–2487.
152. Akalin, A., Garrett-Bakelman, F. E., Kormaksson, M., Busuttil, J., Zhang, L., Khrebtukova, I., Milne, T. A. et al. 2012 Base-pair resolution DNA methylation sequencing reveals profoundly divergent epigenetic landscapes in acute myeloid leukemia. *PLoS Genet.* 8, e1002781.
153. Yamazaki, J., Jelinek, J., Lu, Y., Cesaroni, M., Madzo, J., Neumann, F., He, R. et al. 2015 TET2 Mutations affect non-CpG island DNA

methylation at enhancers and transcription factor-binding sites in chronic myelomonocytic leukemia. *Cancer Res.* 75, 2833–2843.
154. Losman, J. A., Looper, R. E., Koivunen, P., Lee, S., Schneider, R. K., McMahon, C., Cowley, G. S., Root, D. E., Ebert, B. L. and Kaelin, W. G., Jr. 2013. (R)-2-hydroxyglutarate is sufficient to promote leukemogenesis and its effects are reversible. *Science* 339, 1621–1625.
155. Dang, L., White, D. W., Gross, S., Bennett, B. D., Bittinger, M. A., Driggers, E. M., Fantin, V. R. et al. 2009 Cancer-associated IDH1 mutations produce 2-hydroxyglutarate. *Nature* 462, 739–744.
156. Xu, W., Yang, H., Liu, Y., Yang, Y., Wang, P., Kim, S. H., Ito, S. et al. 2011 Oncometabolite 2-hydroxyglutarate is a competitive inhibitor of alpha-ketoglutarate-dependent dioxygenases. *Cancer Cell* 19, 17–30.
157. Ward, P. S., Patel, J., Wise, D. R., Abdel-Wahab, O., Bennett, B. D., Coller, H. A., Cross, J. R. et al. 2010 The common feature of leukemia-associated IDH1 and IDH2 mutations is a neomorphic enzyme activity converting alpha-ketoglutarate to 2-hydroxyglutarate. *Cancer Cell* 17, 225–234.
158. Sasaki, M., Knobbe, C. B., Munger, J. C., Lind, E. F., Brenner, D., Brustle, A., Harris, I. S. et al. 2012 IDH1(R132H) mutation increases murine haematopoietic progenitors and alters epigenetics. *Nature* 488, 656–659.
159. Chaturvedi, A., Araujo Cruz, M. M., Jyotsana, N., Sharma, A., Yun, H., Gorlich, K., Wichmann, M. et al. 2013 Mutant IDH1 promotes leukemogenesis in vivo and can be specifically targeted in human AML. *Blood* 122, 2877–2887.
160. Chen, C., Liu, Y., Lu, C., Cross, J. R., Morris, J. P. t., Shroff, A. S., Ward, P. S., Bradner, J. E., Thompson, C. and Lowe, S. W. 2013 Cancer-associated IDH2 mutants drive an acute myeloid leukemia that is susceptible to Brd4 inhibition. *Genes Dev.* 27, 1974–1985.
161. Mingay, M., Chaturvedi, A., Bilenky, M., Cao, Q., Jackson, L., Hui, T., Moksa, M. et al. 2018 Vitamin C-induced epigenomic remodelling in IDH1 mutant acute myeloid leukaemia. *Leukemia* 32, 11–20.
162. Rampal, R., Alkalin, A., Madzo, J., Vasanthakumar, A., Pronier, E., Patel, J., Li, Y. et al. 2014 DNA hydroxymethylation profiling reveals that WT1 mutations result in loss of TET2 function in acute myeloid leukemia. *Cell Rep.* 9, 1841–1855.
163. Wang, Y., Xiao, M., Chen, X., Chen, L., Xu, Y., Lv, L., Wang, P. et al. 2015 WT1 recruits TET2 to regulate its target gene expression and suppress leukemia cell proliferation. *Mol. Cell* 57, 662–673.
164. Hackanson, B., Robbel, C., Wijermans, P. and Lubbert, M. 2005. In vivo effects of decitabine in myelodysplasia and acute myeloid leukemia: Review of cytogenetic and molecular studies. *Ann. Hematol.* 84(Suppl 1), 32–38.
165. Santos, F. P., Kantarjian, H., Garcia-Manero, G., Issa, J. P. and Ravandi, F. 2010 Decitabine in the treatment of myelodysplastic syndromes. *Expert Rev. Anticancer Ther.* 10, 9–22.
166. Shadduck, R. K., Latsko, J. M., Rossetti, J. M., Haq, B. and Abdulhaq, H. 2007 Recent advances in myelodysplastic syndromes. *Exp. Hematol.* 35, 137–143.

PART III

Vitamin C and Cancer Clinical Studies

CHAPTER SEVEN

Vitamin C and Cancer

AN OVERVIEW OF RECENT CLINICAL TRIALS

Channing Paller, Tami Tamashiro, Thomas Luechtefeld, Amy Gravell, and Mark Levine

CONTENTS

In the mid-twentieth century, it was hypothesized that cancer metastases were due to defective connective tissue, as a consequence of vitamin C deficiency [5]. The hypothesis was expanded with the concept that ascorbic acid might inhibit the spread of metastases by inhibiting hyaluronidase [6], and that high doses of ascorbic acid could be therapeutic. With this background, in the 1970s, improved longevity and symptom relief were observed in advanced cancer patients who received up to 10 g of oral and/or intravenous ascorbate [7–11]. Specifically, Ewan Cameron and Linus Pauling reported a 4.2-fold increase in survival of terminal cancer patients compared to historical controls following supplemental intravenous and oral ascorbate [10]. The findings and conclusions were criticized for multiple reasons, including retrospective study design, population bias, absence of blinding, treatment bias, and lack of pathology confirmation. To replicate these observations, Edward Creagan and Charles Moertel with their colleagues at the Mayo Clinic subsequently conducted two randomized, double-blind, placebo-controlled prospective clinical trials. These trials did not find any clinical benefit from a daily oral dose of 10 g of ascorbate in patients with advanced cancer [12,13], and vitamin C was dismissed as a cancer treatment [14,15]. Subsequent pharmacokinetic studies from the U.S. National Institutes of Health found that intravenous ascorbate can produce more than a 100-fold increase in plasma concentrations compared to the maximum tolerated dose from oral administration [16–20]. In fact, when a higher oral dose of vitamin C (over ~100 mg/day) is consumed, it does not remarkably increase steady-state plasma and tissue vitamin C concentrations due to tightly controlled intestinal uptake, tissue transport and utilization, and renal excretion [2,21]. For example, 200 mg of oral vitamin C daily produces a steady-state plasma concentration of approximately 65–70 μM. Although the maximal plasma concentration achievable with oral vitamin C is transient and less than 250 μM [20], higher, or pharmacologic plasma concentrations of greater than 0.5 mM are attainable from intravenous infusions of ascorbate [4]. Thus, failure of initial rigorous clinical trials to show therapeutic benefit of vitamin C might have been due to limited bioavailability from oral dosing [2]. In addition, the early intervention trials that found no or limited clinical effect of vitamin C may have had control and treatment groups that did not substantially differ from each other in plasma and tissue vitamin C concentrations [22,23]. An increased understanding of the pharmacokinetics of vitamin C when administered orally versus intravenously provided the foundation to reexamine the clinical potential of intravenous ascorbate.

The discovery that intravenous ascorbate is likely more effective than orally administered ascorbate has renewed scientific interest over the past two decades in preclinical studies examining vitamin C as a therapeutic agent for cancer [18,24,25]. Ascorbate at 0.1–100 mM concentration demonstrates cytotoxicity in more than 50 cancer cell lines [26–34]. In many though not all studies, cancer cell death is induced by extracellular, as opposed to intracellular, ascorbate, based on inhibition of cell death by catalase [29,35–37]. Ascorbate treatment with 10 mM concentration caused at least 50% decrease in survival of many cancer cell lines, while healthy cells tolerated a higher ascorbate treatment of 20 mM [29,35–37]. Similarly, pharmacologic ascorbate reduces cell proliferation in tumor cell lines [29,38–40] and in mice tumor xenografts [35] with ascorbate-treatment-caused decreased cell proliferation in colon [41], hepatocellular [42], mesothelioma [43], neuroblastoma [44], prostate [37], and pancreatic [45,46] cell lines. In animal models, pharmacologic ascorbate likewise inhibited tumor growth in liver cancer [39], mesothelioma [43], ovarian cancer [35], pancreatic cancer [38,45,46], prostate cancer [47], and sarcoma [48].

There are a number of caveats to consider when analyzing and interpreting data from cancer cell culture models. The intracellular concentration of ascorbate in cell cultures is virtually zero unless cells are acutely isolated [49]. Even when ascorbate is added to the extracellular cell culture media, the intracellular ascorbate concentration depends on various factors, such as oxidation of extracellular ascorbate in media, replenishment of oxidized ascorbate, incubation timing, dehydroascorbic acid transport, ascorbate and facilitated diffusion glucose transporters, and transporter gene copy number [49]. Additionally, it is not appropriate to base dosing decisions solely on *in vitro* and animal data [49], making clinical translation difficult.

ANTICANCER MECHANISMS OF ASCORBATE

Anticancer mechanisms of vitamin C depend on a wide variety of factors, including cancer type and a tumor's dependency on specific pathways. There are multiple cancer vulnerabilities that can be manipulated by pharmacologic ascorbate: via generation of extracellular hydrogen peroxide, targeting extracellular and intracellular redox imbalance, which can be further modulated by increasing labile iron in cells, increasing dehydroascorbic acid uptake, or decreasing NAD; targeting epigenetic regulators; and targeting signaling of hypoxia-inducible factor 1α [50–52]. Table 7.1 lists some pivotal studies in analyses of vitamin C in cancer patients over the past 60 or more years.

While vitamin C concentrations in cancer patients are often deficient [53–55], pharmacologic ascorbate is not simply correcting deficiency. Instead, pharmacologic ascorbate is targeting induction of oxidative stress in cancer cells [29,35,36] and delivery of ascorbate in tumors for optimal cofactor activity of DNA demethylation and hypoxia inducible factor hydroxylation [56]. Extracellular pharmacologic ascorbate (but not physiologic ascorbate) produces high levels of extracellular hydrogen peroxide [4,29,36,45,57,58], causing oxidative stress that is selectively toxic to different types of cancer cells and not to normal cells of the same tissue type [29,35–37,39,45,46,59–66]. Because hydrogen peroxide diffuses into cells, extracellular hydrogen peroxide induces both extracellular and intracellular oxidative damage. Pharmacologic ascorbate generates hydrogen peroxide in animal studies [35,36], and cases of oxidative damage (oxidative hemolysis) in patients with glucose 6-phosphate dehydrogenase deficiency after receiving pharmacologic ascorbate [67,68] implicate that hydrogen peroxide formation occurs in humans [49]. It is likely that many different cellular pathways are involved in repairing cell damage in response to hydrogen peroxide formation. Despite common sensitivity to hydrogen peroxide, repair pathways that are compromised in cancer cells are likely to vary and to be cell type specific [2]. Nevertheless, damage to cancer cells from extracellular hydrogen

TABLE 7.1
Evolution of ascorbic acid studies in cancer[a]

Cameron/Pauling Studies	Mayo Clinic Studies	Vitamin C Pharmacokinetics and Early Phase Clinical Trials	Studies on H_2O_2 Mechanisms of Anticancer Activity	Studies on Epigenetic Mechanisms of Anticancer Activity	Studies of Metabolic and Hypoxia-Inducible Factor (HIF)-1α-Dependent Mechanisms of Anticancer Activity
Ewan Cameron, joined by Linus Pauling, described retrospectively and in case reports that patients with advanced cancer had survival benefit and symptomatic relief using high-dose ascorbate (10 g/day intravenous [IV] followed by oral) [8,10,11]	Two rigorous double-blind, placebo-controlled prospective trials performed at the Mayo Clinic using the same dose of ascorbate, but orally only, failed to confirm these results, and oral ascorbate was dismissed as an anticancer agent [12,13,15].	Oral ascorbic acid over an 80-fold dose range was found to produce plasma concentrations that were tightly regulated by gastrointestinal absorption, but IV administration bypassed this control until the kidney restored homeostasis. Maximum tolerated doses of oral (18 g daily) ascorbate produced plasma concentrations of 100–200 mM. IV ascorbate was found to produce plasma levels hundreds of times higher than those produced by the maximum tolerated dose of oral ascorbate [17–20]. Early phase clinical trials indicate that IV ascorbate at 1 g/kg over 1.5–2 hours two to three times weekly is well tolerated and may enhance chemosensitivity as well as decrease chemotherapy-related side effects [66,99,101,102,128,142,145,173].	Plasma concentrations achieved by IV dosing found to act as a prodrug for hydrogen peroxide (H_2O_2) in the extracellular space. Pharmacologic, but not physiologic, ascorbic acid was selectively toxic to cancer cells *in vitro* and *in vivo* [29,35,36,39, 46,66,161,162].	Ascorbate functions as a cofactor and increases the activity of the ten-eleven translocation (TET) enzymes causing DNA demethylation. This function has been found to result in the reexpression of tumor-suppressor genes in cancer cells, promotion of stem cell differentiation and inhibition of leukemogenesis, and enhancement of DNA. Methyltransferase inhibitor (DNMTi) induced immune signals via increased expression of endogenous retrovirus transcripts [54,77–80].	Metabolic toxicity to breast and neuroblastoma cancer cells occurs from NAD depletion [73,174]. Increased uptake of oxidized vitamin C, dehydroascorbic acid (DHA) may lead to inactivation of glyceraldehyde 3-phosphate dehydrogenase (GAPDH) and cell death in highly glycolytic *KRAS* or *BRAF* mutant colon cancer cells [71], and erlotinib-resistant lung adenocarcinoma cells [74]. Ascorbic acid may downregulate HIF 1α expression in leukemia, endometrial, and colorectal cancer cells [50–52].

[a] Reproduced and revised from Shenoy N. et al. *Cancer Cell* 34, 700–706.

peroxide appears to be mediated through multiple pathways that are initiated either outside or inside cells, meaning that pharmacologic ascorbate is a pluripotent agent, or promiscuous agent [69].

While pharmacologic ascorbate appears to have cytotoxic effects on many cancer cells through hydrogen-peroxide-mediated pro-oxidant damage, in a subset of cancer cells, additional related mechanisms have been described. Cytotoxicity may be due to oxidation of ascorbate into an unstable metabolite and reversible oxidized form of ascorbate, dehydroascorbic acid [70]. Tumor cells internally reduce dehydroascorbic acid to ascorbate-triggering glutathione scavenging, inducing oxidative stress, inactivating glyceraldehyde 3-phosphate dehydrogenase, inhibiting glycolytic flux, and ultimately leading to an energy crisis leading to cell death [71,72]. For example, cultured human colorectal cancer cells with *KRAS* or *BRAF* mutations were selectively killed by pharmacologic ascorbate by depletion of intracellular glutathione. This is followed by inactivation of glyceraldehyde 3-phosphate dehydrogenase, leading to inhibition of glycolysis and death in cancer cells highly dependent on glycolysis [71]. Pharmacologic ascorbate can also induce metabolic stress by depletion of NAD in several cancer cell lines [73,74].

An independent mechanism of the antitumor action of vitamin C involves an epigenetic DNA demethylation from activated ten-eleven translocation (TET) enzymes. Ascorbate increases TET enzyme activity, resulting in DNA demethylation and increased hydroxymethylation [75,76]. This can then cause a variety of actions: reexpression of tumor suppressor genes in cancer cells, inhibition of leukemogenesis, and expression of endogenous retrovirus transcripts that enhance immune signals induced by DNA methyltransferase inhibitor [54,77–80]. Ascorbate facilitates TET-mediated DNA demethylation by binding to the catalytic domain of TET enzymes [75,79], which reverses hypermethylation in tumors and activates tumor suppressor genes and mechanisms [79,81]. The *TET2* gene is frequently mutated in hematopoietic malignancies and associated with DNA hypermethylation, heightened risk of myelodysplastic syndrome progression, and poor acute myeloid leukemia prognosis [82]. Vitamin C mimics restoration of *TET2* in leukemic stem cells and enhances poly-ADP ribose polymerase inhibition in suppressing leukemia progression [78]. The ability of vitamin C to enhance TET function makes it a potential therapy for TET-associated malignancies, and future clinical trials should include high-dose vitamin C ancillary to standard chemotherapy/demethylating treatment, especially for *TET2*-deficient tumors [78]. It is currently unclear whether physiologic doses of vitamin C result in DNA demethylation, but it is hypothesized that vitamin C cofactor functions, such as DNA demethylation, would be similar or higher in pharmacologic doses [49]. Pharmacologic ascorbate also needs to be further investigated as a therapeutic treatment, in particular for malignancies with aberrant hypermethylation, including chronic myelomonocytic leukemia, myelodysplastic syndrome, acute myelogenous leukemia (*TET2* aberration with either *IDH1* or *IDH2* aberration), clear-cell renal cell carcinoma, and succinate dehydrogenase mutated paraganglioma [83–90].

Another mechanism of pharmacologic ascorbate action may be mediated enzymatically via hydroxylation of hypoxia inducible factor enzymes, rendering them inactive [50–52]. This mechanism is discussed in detail separately (see Chapter 4).

USE OF THE CLINICALTRIALS.GOV DATABASE

ClinicalTrials.gov, which is managed by the U.S. National Library of Medicine, is a database of publicly and privately funded clinical studies conducted in 206 countries. Information about each clinical trial is provided by the trial's sponsor or principal investigator. The ClinicalTrials.gov website became publicly available in February 2000, and the results database on the site containing information on study participants, adverse events, and outcomes became available in September 2008. Congress passed the Food and Drug Administration Amendments Act of 2007 that expanded clinical trial registration requirements to ClinicalTrials.gov, which required additional types of trials to be registered and required additional trial information, such as summary results. Due to the staggered implementation of registration requirements for ClinicalTrials.gov over the past two decades, not all clinical trials are included in the database, and information on trials is incomplete. Despite these limitations, ClinicalTrials.gov is still the largest trial registry in the world.

DATA EXTRACTION METHOD

Clinical trials included in the ClinicalTrials.gov database were filtered with the condition key word "cancer" and intervention key words "vitamin C" OR "ascorbate" OR "ascorbic acid" on November 29, 2018, to identify clinical trials in which cancer patients were treated with vitamin C as a single agent or in combination with standard therapies. The resulting clinical trials were extracted and imported into a database, which is available online at https://sysrev.com/p/6737. Trials were manually reviewed to confirm that they were currently or previously active and that vitamin C was indeed administered to patients. This review of data was completed by January 1, 2019. A total of 39 trials were excluded from analysis. Trials that were not included in the original ClinicalTrials.gov search were classified as suspended, not yet recruiting, no longer available, withheld, temporarily not available, excluded, withdrawn, or terminated. In addition, five trials were also excluded from analysis because they used vitamin C as a dietary supplement and not as an investigational treatment. The remaining 42 trials, which were all associated with a National Clinical Trial (NCT) identifying number, were analyzed to generate all figures. Some fields (e.g., concurrent treatment dose) for each clinical trial similarly needed to be manually extracted from the trial description in ClinicalTrials.gov, and then mapped to common names. While the mesh term "vitamins" was initially used to identify oral vitamin C trials, the authors manually identified oral vitamin C trials [91].

LITERATURE SEARCH TO SUPPLEMENT CLINICALTRIALS.GOV

A 2018 review article included three tables listing intravenous vitamin C cancer trials based on a literature search in the PubMed, MEDLINE, and Cochrane databases [92]. These trials were compared with those identified from ClinicalTrials.gov, and a total of 11 trials not identified in the ClinicalTrials.gov search were included in Tables 7.2–7.4 to more fully capture cancer clinical trials and associated published results that utilized vitamin C as a therapeutic treatment. Although these additional 11 trials were included in Tables 7.2–7.4, they were not added to the ClinicalTrials.gov data extraction and are not reflected in the figures of this chapter due to missing trial identifiers, such as NCT numbers. Despite using this combined literature-based and ClinicalTrials.gov-based approach to identify vitamin C cancer trials, our final repository still does not capture all vitamin C cancer trials, especially those prior to 2000.

SOURCES OF VITAMIN C

Intravenous ascorbate is commonly derived from corn but may also be derived from beets, sago palm, and other foods. Corn-free ascorbate (usually derived from sugar beets) is indicated for patients who have a corn allergy. Buffered vitamin C, using magnesium, sodium, calcium, or potassium ascorbate, has a lower acid content than nonbuffered ascorbic acid [93]. Ascorbate should never be administered in its pure acid form but rather in preparations that are pH adjusted. Data are very limited about differences between ascorbate sources regarding clinical use. It is unknown whether trace materials from different ascorbate sources have clinical relevance. It is likely that most trace compounds are removed during manufacturing and purification [94].

CHARACTERISTICS OF VITAMIN C CANCER CLINICAL TRIALS

Out of a total of 42 vitamin C trials identified in ClinicalTrials.gov, 19 were completed (16 intravenous and 3 oral ascorbate), 6 were active but not recruiting participants (5 intravenous and 1 oral ascorbate), and 17 were actively recruiting (14 intravenous and 3 oral ascorbate). In the past 3 years, the number of cancer vitamin C trials doubled compared to the prior 16 years (Figure 7.1a and b). Much of that increase can be attributed to high-dose intravenous vitamin C trials. Intravenous administration of vitamin C is required for proposed anticancer activity, to provide higher peak plasma concentration, and to enhance diffusion of vitamin C into the hypoxic core of solid tumors [95]. There is some evidence that ascorbate accumulates in solid tumors at higher concentrations than surrounding tissue [96–98], suggesting that cancer cells are particularly affected by vitamin C treatment. Many high-dose intravenous vitamin C trials are not completed and are still accruing participants (Table 7.2), so data to inform clinical practice and to develop future clinical trials are currently limited.

TABLE 7.2

High-dose and pharmacologic intravenous vitamin C clinical trials (greater than 1 g daily intravenous dose)[a]

National Clinical Trial (NCT) Number	Reference	n	Patient Diagnosis	Trial Design/ Phase	Intravenous (IV) Ascorbate (Vitamin C) Dose and Frequency	Concurrent Treatment Dose	Toxicity	Reported Outcomes/ Conclusions
NCT02420314	[94] Schoenfeld, 2017	57 anticipated	Late-stage non-small cell lung cancer	Single arm Phase 2	IV ascorbate (75 g twice per week for 21-day cycle, up to four cycles)	Carboplatin (AUC = 6, once every cycle [21 days], up to four cycles) Paclitaxel (200 mg/m^2 once every cycle [21 days], up to four cycles)	Interim analysis (n = 14): No grade 3 or 4 toxicities related to ascorbate occurred at interim analysis.	At interim analysis (n = 14): Disease control rate 93%; objective response rate 29%; imaging-confirmed partial responses to therapy (n = 4), stable disease (n = 9), and disease progression (n = 1).
NCT02905578	None given	65 anticipated	Metastatic pancreatic cancer	Randomized Phase 2	Experimental arm: IV ascorbate (75 g three times per week for 4-week cycle; cycles continue until disease progression)	Experimental arm: Gemcitabine (1000 mg/m^2 once weekly for 3 weeks in 4-week cycle; cycles continue until disease progression) Nab-paclitaxel (125 mg/m^2 once weekly for 3 weeks in 4-week cycle; cycles continue until disease progression)	Not available	Not available
					Control arm: No IV ascorbate	Control arm: Gemcitabine (1000 mg/m^2 once weekly for 3 weeks in 4-week cycle; cycles continue until disease progression) Nab-paclitaxel (125 mg/m^2 once weekly for 3 weeks in 4-week cycle; cycles continue until disease progression)		

(Continued)

TABLE 7.2 *(Continued)*

High-dose and pharmacologic intravenous vitamin C clinical trials (greater than 1 g daily intravenous dose)[a]

National Clinical Trial (NCT) Number	Reference	n	Patient Diagnosis	Trial Design/ Phase	Intravenous (IV) Ascorbate (Vitamin C) Dose and Frequency	Concurrent Treatment Dose	Toxicity	Reported Outcomes/ Conclusions
NCT02344355	None given	90 anticipated	Glioblastoma multiforme	Single arm Phase 2	Combined radiation and chemotherapy phase: IV ascorbate (87.5 g three times per week)	Combined radiation and chemotherapy phase: Conformal radiation therapy (61.2 Gy in 34 fractions administered daily Monday to Friday); oral temozolomide (75 mg/m^2; 7 days per week up to 49 days during radiation therapy)	Not available	Not available
					Adjuvant chemotherapy phase: IV ascorbate (87.5 g twice per week for six 28-day cycles)	Starting 1 month after radiation therapy: Oral temozolomide (first cycle: 150 mg/m^2 once daily for 5 days in 28-day cycle; if first cycle is well tolerated, then dose escalated to 200 mg/m^2 for cycles two to six)		

(Continued)

TABLE 7.2 *(Continued)*

High-dose and pharmacologic intravenous vitamin C clinical trials (greater than 1 g daily intravenous dose)[a]

National Clinical Trial (NCT) Number	Reference	n	Patient Diagnosis	Trial Design/ Phase	Intravenous (IV) Ascorbate (Vitamin C) Dose and Frequency	Concurrent Treatment Dose	Toxicity	Reported Outcomes/ Conclusions
NCT02905591	None given	46 anticipated	Non-small cell lung cancer	Single arm Phase 2	IV ascorbate (75 g three times per week for 6–7 weeks)	Conformal radiation therapy (60 Gy in two fractions; one fraction delivered daily for 30 fractions in 30 days) Paclitaxel (45 mg/m^2 administered weekly for ~6–7 weeks; followed by two extra cycles of consolidation chemotherapy) Carboplatin (AUC = 2, administered weekly for ~6–7 weeks; followed by two extra cycles of consolidation chemotherapy)	Not available	Not available
NCT03433781	None given	18 anticipated	Intermediate or high-risk myelodysplastic syndrome with TET2 mutations	Single arm Phase 1/2	IV ascorbate (50 g daily for 5 days every 4 weeks)	None	Not available	Not available
NCT03410030	None given	36 anticipated	Metastatic pancreatic cancer	Single arm Phase 1/2	No specific intravenous ascorbate dosing provided, but "high-dose" $\geq$20 mM plasma concentration referenced	Nanoparticle paclitaxel protein bound, Gemcitabine, Cisplatin (no exact dosing provided)	Not available	Not available

(Continued)

TABLE 7.2 (*Continued*)

High-dose and pharmacologic intravenous vitamin C clinical trials (greater than 1 g daily intravenous dose)[a]

National Clinical Trial (NCT) Number	Reference	n	Patient Diagnosis	Trial Design/ Phase	Intravenous (IV) Ascorbate (Vitamin C) Dose and Frequency	Concurrent Treatment Dose	Toxicity	Reported Outcomes/ Conclusions
NCT02516670	None given	69 anticipated	Metastatic prostate cancer	Randomized Phase 2	Experimental arm: IV ascorbate (1 g/kg twice per week (ideally 3–4 days apart) for 3-week cycle up to eight cycles) For first cycle only: First three doses are titrated 0.25, 0.5, and 0.75 g/kg in 1000 mL of water	Experimental arm: Docetaxel (75 mg/m^2 every 3 weeks on day 1 of the cycle for up to eight cycles)	Not available	Not available
					Placebo arm: No IV ascorbate treatment	Placebo arm: Docetaxel (75 mg/m^2 every 3 weeks on day 1 of the cycle for up to eight cycles); placebo (1000 mL of normal saline twice per week for 3-week cycle up to eight cycles)		
NCT03334409	None given	91 anticipated	Metastatic or unresectable clear cell renal cell carcinoma	Randomized Phase 2	Experimental arm: IV ascorbate (1 g/kg three times per week up to 40 weeks)	Experimental arm: Pazopanib (800 mg daily on days 1–28 in 28-day cycle up to 10 cycles)	Not available	Not available
					Control arm: No intravenous ascorbate	Control arm: Pazopanib (800 mg daily on days 1–28 in 28-day cycle up to 10 cycles)		

(*Continued*)

TABLE 7.2 (*Continued*)

High-dose and pharmacologic intravenous vitamin C clinical trials (greater than 1 g daily intravenous dose)[a]

National Clinical Trial (NCT) Number	Reference	n	Patient Diagnosis	Trial Design/ Phase	Intravenous (IV) Ascorbate (Vitamin C) Dose and Frequency	Concurrent Treatment Dose	Toxicity	Reported Outcomes/ Conclusions
NCT03146962	None given	50 anticipated	Cohort A: Early stage or locally advanced colorectal, lung, or pancreatic cancer, who are eligible for resection Cohort B: Inoperable, metastatic, KRAS or BRAF mutant colorectal, lung, or pancreatic cancer	Single arm Phase 2	Intravenous ascorbate (1.25 g/kg for 4 days per week for 2–4 consecutive weeks [cohort A] or up to 6 months [cohort B])	None	Not available	Not available
NCT03015675	None given	200 anticipated	Metastatic gastric cancer	Randomized Phase 3	Experimental arm: IV ascorbate (20 g/day on days 1–3 every 2 weeks)	Experimental arm: mFOLOX6: - Oxaliplatin 85 mg/m^2 day 1 concurrent with - Leucovorin 400 mg/m^2, followed by - Bolus 5FU 400 mg/m^2, followed by - 5FU infusion 2400 mg/m^2 over 46 hours, every 2 weeks	Not available	Not available

(Continued)

TABLE 7.2 (*Continued*)

High-dose and pharmacologic intravenous vitamin C clinical trials (greater than 1 g daily intravenous dose)[a]

National Clinical Trial (NCT) Number	Reference	n	Patient Diagnosis	Trial Design/ Phase	Intravenous (IV) Ascorbate (Vitamin C) Dose and Frequency	Concurrent Treatment Dose	Toxicity	Reported Outcomes/ Conclusions
					Control arm: No intravenous ascorbate	Control arm: mFOLOX6: - Oxaliplatin 85 mg/m^2 day 1 concurrent with - Leucovorin 400 mg/m^2, followed by - Bolus 5FU 400 mg/m^2, followed by - 5FU infusion 2400 mg/m^2 over 46 hours, every 2 weeks		
NCT02969681	None given	428 anticipated	Metastatic colorectal cancer	Randomized Phase 3	Experimental arm: Ascorbic acid (1.5 g/kg/day on day 1–3 every 2 weeks)	Experimental arm: mFOLFOX6: - Oxaliplatin 85 mg/m^2 day 1 concurrent with - Leucovorin 400 mg/m^2, followed by - Bolus 5FU 400 mg/m^2, followed by - 5FU infusion 2400 mg/m^2 over 46 hours, every 2 weeks With or without, bevacizumab (5 mg/kg every 2 weeks)	Not available	Not available

(Continued)

TABLE 7.2 *(Continued)*

High-dose and pharmacologic intravenous vitamin C clinical trials (greater than 1 g daily intravenous dose)[a]

National Clinical Trial (NCT) Number	Reference	n	Patient Diagnosis	Trial Design/ Phase	Intravenous (IV) Ascorbate (Vitamin C) Dose and Frequency	Concurrent Treatment Dose	Toxicity	Reported Outcomes/ Conclusions
					Control arm: No IV ascorbate	Control arm: mFOLFOX6: - Oxaliplatin 85 mg/m^2 day 1 concurrent with - Leucovorin 400 mg/m^2, followed by - Bolus 5FU 400 mg/m^2, followed by - 5FU infusion 2400 mg/m^2 over 46 hours, every 2 weeks With or without, bevacizumab (5 mg/kg every 2 weeks)		
NCT03613727	None given	60 anticipated	Hodgkin lymphoma, lymphoid leukemia, multiple myeloma, myeloid leukemia, monocytic leukemia, non-Hodgkin lymphoma, myelodysplasia	Single arm Phase 2	IV ascorbate (16.7 mg/kg three times a day [every 8 hours] beginning on posttransplant day 1 through day 14 After completion of IV ascorbate, oral vitamin C is administered (500 mg twice each day until 6 months after transplant)	None	Not available	Not available

(Continued)

TABLE 7.2 (*Continued*)

High-dose and pharmacologic intravenous vitamin C clinical trials (greater than 1 g daily intravenous dose)[a]

National Clinical Trial (NCT) Number	Reference	n	Patient Diagnosis	Trial Design/ Phase	Intravenous (IV) Ascorbate (Vitamin C) Dose and Frequency	Concurrent Treatment Dose	Toxicity	Reported Outcomes/ Conclusions
NCT03418038	None given	151 anticipated	Relapsed or refractory lymphoma	Randomized Phase 2	Experimental arm A: IV ascorbate (1 g/kg (maximum 100 g) on days 1, 3, 5, 8, 10, 12, 15, 17, and 19	No doses provided Experimental arm A: IV rituximab, ifosfamide, carboplatin, and etoposide on days 1–3 Patients who achieve minor response (MR) or stable disease (SD) after two courses may receive IV or oral rituximab, cisplatin, cytarabine, and dexamethasone Treatment repeats every 21 days up to four courses	Not available	Not available
					Placebo/active comparator arm B: No IV ascorbate treatment	No doses provided Placebo arm B: Normal saline on days 1, 3, 5, 8, 10, 12, 15, 17, and 19 IV rituximab, ifosfamide, carboplatin, and etoposide on days 1–3 Patients who achieve MR or SD after two courses may receive IV or oral rituximab, cisplatin, cytarabine, and dexamethasone Treatment repeats every 21 days up to four courses		

(*Continued*)

TABLE 7.2 *(Continued)*

High-dose and pharmacologic intravenous vitamin C clinical trials (greater than 1 g daily intravenous dose)[a]

National Clinical Trial (NCT) Number	Reference	n	Patient Diagnosis	Trial Design/ Phase	Intravenous (IV) Ascorbate (Vitamin C) Dose and Frequency	Concurrent Treatment Dose	Toxicity	Reported Outcomes/ Conclusions
					Experimental arm C: IV ascorbate (1 g/kg [maximum 100 g] on days 1, 3, 5, 8, 10, 12, 15, 17, and 19)	No doses provided Experimental arm C: ifosfamide, carboplatin, and etoposide; or cisplatin, cytarabine, and dexamethasone; or gemcitabine hydrochloride, dexamethasone, and cisplatin; or gemcitabine hydrochloride and oxaliplatin; or oxaliplatin, cytarabine, and dexamethasone Treatment repeats every 21 days for up to four courses Patients who achieve MR or SD after two courses may switch to an alternative chemotherapy regimen		
NCT03468075	None given	29 anticipated	Advanced, unresectable, or metastatic soft tissue or bone sarcoma	Single arm Phase 2	IV ascorbate (75 g on days 1, 2, 8, 9, 15, and 16 of a 28-day cycle)	No dose provided Gemcitabine administered on days 1, 8, and 15, after the infusion of ascorbate	Not available	Not available

(Continued)

TABLE 7.2 (*Continued*)

High-dose and pharmacologic intravenous vitamin C clinical trials (greater than 1 g daily intravenous dose)[a]

National Clinical Trial (NCT) Number	Reference	n	Patient Diagnosis	Trial Design/ Phase	Intravenous (IV) Ascorbate (Vitamin C) Dose and Frequency	Concurrent Treatment Dose	Toxicity	Reported Outcomes/ Conclusions
NCT01752491	[94] Schoenfeld, 2017	13 enrolled	Glioblastoma multiforme	Single arm Phase 1	Radiation phase: IV ascorbate (doses ranging from 15–125 g, three times per week for 7 weeks with a 20 mM target plasma concentration) Adjuvant phase: IV ascorbate (dose escalation until ≥20 mM plasma concentration reached, twice per week for ~28 weeks)	Radiation phase: Radiation (61.2 Gy in 34 fractions); temozolomide (75 mg/m^2 daily until radiation is completed for a maximum of 49 days) Adjuvant phase: Temozolomide (150 mg/m^2 and then one dose escalation to 200 mg/m^2 if no toxicity in cycle 1; daily for 5 days followed by 23 days of rest for six, 28-day cycles)	Ascorbate was safe and well tolerated (n = 13) with minimal grades 3 and 4 toxicities. No serious adverse events were ascribed to ascorbate treatment Nonhematologic radiation phase toxicity: grades 2 and 3 fatigue and nausea; grade 2 infection; grade 3 vomiting Adjuvant phase toxicity: grade 2 fatigue and nausea; grade 1 vomiting; grade 3 leukopenia; and grade 3 neutropenia	Average progression-free survival 13.3 months (n = 11 analyzable patients); average overall survival 21.5 months (n = 11 analyzable patients).

(*Continued*)

TABLE 7.2 (*Continued*)

High-dose and pharmacologic intravenous vitamin C clinical trials (greater than 1 g daily intravenous dose)[a]

National Clinical Trial (NCT) Number	Reference	n	Patient Diagnosis	Trial Design/ Phase	Intravenous (IV) Ascorbate (Vitamin C) Dose and Frequency	Concurrent Treatment Dose	Toxicity	Reported Outcomes/ Conclusions
NCT01852890	[121] Alexander, 2018	16 enrolled	Pancreatic cancer	Single arm Phase 1	IV ascorbate (50, 75, or 100 g daily during radiation therapy for 5–6 weeks; escalating dose based on tolerance)	Gemcitabine (600 mg/m^2 once weekly for 6 weeks) Intensity-modulated radiotherapy (either 50 Gy in 25 fractions or 50.4 Gy in 28 fractions as determined most appropriate by the treating radiation oncologist; one fraction per day Monday to Friday)	No adverse events during the ascorbate infusion with combination therapy or immediately thereafter Three adverse events attributable to ascorbate treatment: dry mouth, thirst, and transient blood pressure elevation One dose-limiting toxicity occurred at 75 g ascorbate dose: grade 3 noncardiac chest pain (patient rechallenged without incident and continued in the trial) Grades 3 and 4 hematologic toxicities were consistent with gemcitabine and radiotherapy	First-in-human phase 1 trial that infused a high dose of IV ascorbate "pharmacologic ascorbate" during the radiotherapy "beam on." Pharmacologic ascorbate in combination with gemcitabine and radiotherapy for locally advanced pancreatic cancer is safe and well tolerated with suggested efficacy. Pharmacologic ascorbate increased the median overall survival compared with the institutional average (21.7 versus 12.7 months) and compared with the E4201 trial (21.7 versus 11.1 months) (n = 14 evaluable patients). Maximum tolerated dose of ascorbate infusion: 100 g. Recommended phase 2 dose of ascorbate infusion: 75 g.

(**Continued**)

TABLE 7.2 (*Continued*)

High-dose and pharmacologic intravenous vitamin C clinical trials (greater than 1 g daily intravenous dose)[a]

National Clinical Trial (NCT) Number	Reference	n	Patient Diagnosis	Trial Design/ Phase	Intravenous (IV) Ascorbate (Vitamin C) Dose and Frequency	Concurrent Treatment Dose	Toxicity	Reported Outcomes/ Conclusions
NCT02896907	None given	8 enrolled	Advanced or recurrent unresectable pancreatic cancer	Single arm Early phase 1	IV ascorbate (starting dose of 75 g and subsequent doses of 100 g over 2 hours on days 3, 5, 8, 10, and 12; dosing repeated every 14 days)	No doses provided FOLFIRINOX: oxaliplatin, irinotecan hydrochloride, leucovorin calcium, and fluorouracil continuously over 46 hours on day 1; dosing repeated every 14 days	Not available	Not available
NCT02833701	None given	9 enrolled	Recurrent high-grade glioma	Single arm Phase 1	IV ascorbate (25–100 g) over 90–120 minutes three times per week (at least 24 hours apart) every 28 days up to 12 courses	No dose provided IV bevacizumab over 30–90 minutes on days 1 and 15 every 28 days for up to 12 courses	Not available	Not available

(*Continued*)

TABLE 7.2 (*Continued*)

High-dose and pharmacologic intravenous vitamin C clinical trials (greater than 1 g daily intravenous dose)[a]

National Clinical Trial (NCT) Number	Reference	n	Patient Diagnosis	Trial Design/ Phase	Intravenous (IV) Ascorbate (Vitamin C) Dose and Frequency	Concurrent Treatment Dose	Toxicity	Reported Outcomes/ Conclusions
NCT01905150	[154] Bruckner, 2017	30 enrolled	Advanced pancreatic cancer	Randomized Phase 2	Experimental arm: IV ascorbate (75–100 g administered on average one to two times per week throughout the study)	No doses provided Experimental arm: G-FLIP treatment followed by G-FLIP-DM treatment G-FLIP treatment: Gemcitabine, fluorouracil, leucovorin, irinotecan, and oxaliplatin on days 1 and 2 of each cycle for 2 weeks G-FLIP-DM treatment includes G-FLIP + low doses of docetaxel and mitomycin C	Five out of 26 patients analyzed (19%) had grade 3 neutropenia. One out of 26 patients (4%) had grade 3 anemia and thrombocytopenia. No limiting gastrointestinal toxicity, weight loss, or coagulopathy due to G-FLIP treatment	High-dose ascorbate combined with G-FLIP chemotherapy is safe and well tolerated for high-risk patients with pancreatic cancer and may avoid 20%–40% rates of severe toxicities.
					Control arm: No IV ascorbate	No doses provided Experimental arm: G-FLIP then followed by G-FLIP-DM treatment G-FLIP includes gemcitabine, fluorouracil, leucovorin, irinotecan, and oxaliplatin on days 1 and 2 of each cycle for 2 weeks G-FLIP-DM treatment includes G-FLIP + low doses of docetaxel and mitomycin C		

(Continued)

TABLE 7.2 (*Continued*)

High-dose and pharmacologic intravenous vitamin C clinical trials (greater than 1 g daily intravenous dose)[a]

National Clinical Trial (NCT) Number	Reference	n	Patient Diagnosis	Trial Design/ Phase	Intravenous (IV) Ascorbate (Vitamin C) Dose and Frequency	Concurrent Treatment Dose	Toxicity	Reported Outcomes/ Conclusions
NCT00441207	[120] Stephenson, 2013	18 enrolled	Advanced solid tumors	Single arm Phase 1	IV ascorbate (five cohorts treated with 30, 50, 70, 90, or 110 g/m^2 for 4 consecutive days per week for 4 weeks)	None	All doses well tolerated Grade 3 toxicities: hypernatremia, hypokalemia, and headache Grade 4 toxicities: hypernatremia	No objective antitumor response in patients (n = 16). Three patients had stable disease, 13 had progressive disease. Cmax and AUC values increased proportionately with ascorbate dose between 0 and 70 g/m^2 but reached maximal values at 70 g/m^2 (49 mM and 220 h mM, respectively). Recommended IV ascorbate dose is 70–80 g/m^2.
None given	[161] Kawada, 2014	3 enrolled	Relapsed B-cell non-Hodgkin lymphoma	Single arm Phase 1	Test dose of IV ascorbate (15 g) administered on day 7 and then 75 or 100 g administered on days 9, 11, 14, 16, and 18 during the second course of the CHASER regimen	CHASER regimen: Rituximab (375 mg/m^2), cyclophosphamide (1200 mg/m^2), cytarabine (2 g/m^2), etoposide (100 mg/m^2), and dexamethasone (40 mg)	No adverse reactions attributed to IV ascorbate treatment. CHASER regimen attributed with hematologic toxicities: grade 3 neutropenia, anemia, and thrombocytopenia	IV ascorbate dose of 75 g was safe and sufficient to achieve >15 mM serum concentration.

(*Continued*)

TABLE 7.2 (*Continued*)

High-dose and pharmacologic intravenous vitamin C clinical trials (greater than 1 g daily intravenous dose)[a]

National Clinical Trial (NCT) Number	Reference	n	Patient Diagnosis	Trial Design/ Phase	Intravenous (IV) Ascorbate (Vitamin C) Dose and Frequency	Concurrent Treatment Dose	Toxicity	Reported Outcomes/ Conclusions
NCT01049880	[95] Welsh, 2013	15 enrolled	Unresectable, metastatic, or recurrent pancreatic cancer	Single arm Phase 1	IV ascorbate (15–125 g twice weekly with escalating doses to achieve a targeted plasma level of ≥20 mM)	Gemcitabine (1000 mg/m^2 weekly for 3 weeks)	Among nine evaluable patients, toxicities possibly attributable to ascorbate included nausea (n = 6), diarrhea (n = 4), and thirst/dry mouth (n = 4). No dose-limiting toxicities or serious adverse events occurred. Treating metastatic pancreatic cancer patients with pharmacologic ascorbate in combination with gemcitabine should be safe and well tolerated	Mean survival time of patients completing at least 8 weeks of therapy was 13 months. Two out of 11 treated patients discontinued treatment within the first month due to progressive disease. Six of nine evaluable patients maintained or improved performance status. Mean progression-free survival was 26 weeks, and overall survival was 12 months (n = 9). High-dose IV ascorbate administered concurrently with gemcitabine is well tolerated.

(*Continued*)

TABLE 7.2 (*Continued*)

High-dose and pharmacologic intravenous vitamin C clinical trials (greater than 1 g daily intravenous dose)[a]

National Clinical Trial (NCT) Number	Reference	n	Patient Diagnosis	Trial Design/ Phase	Intravenous (IV) Ascorbate (Vitamin C) Dose and Frequency	Concurrent Treatment Dose	Toxicity	Reported Outcomes/ Conclusions
NCT00954525	[92] Monti, 2012	14 enrolled	Metastatic pancreatic cancer	Single arm Phase 1	IV ascorbate (50, 75, or 100 g per infusion [three cohorts] three times a week for 8 weeks)	Gemcitabine (1000 mg/m^2 on day 1 then once weekly for 7 weeks followed by a 1-week rest) and erlotinib (100 mg/day for 8 weeks)	Fifteen nonserious adverse events and 8 serious adverse events related to progression of disease or treatment with gemcitabine or erlotinib. Two grade 4 pulmonary embolism adverse events attributable to underlying cancer. No increased toxicity with the addition of ascorbate to gemcitabine and erlotinib in pancreatic cancer patients	Seven of the nine patients (who completed treatment) had stable disease, while the other two had progressive disease. Eight out of nine patients had tumor shrinkage after only 8 weeks of treatment. Estimated mean progression-free survival of 12.7 weeks and overall survival of 6 months. Five out of 14 patients did not complete the study; three out of five of these patients died from rapid disease progression. Overall mean survival was 182 days.

(*Continued*)

TABLE 7.2 (*Continued*)

High-dose and pharmacologic intravenous vitamin C clinical trials (greater than 1 g daily intravenous dose)[a]

National Clinical Trial (NCT) Number	Reference	n	Patient Diagnosis	Trial Design/ Phase	Intravenous (IV) Ascorbate (Vitamin C) Dose and Frequency	Concurrent Treatment Dose	Toxicity	Reported Outcomes/ Conclusions
NCT00228319	[62] Ma, 2014	27 enrolled	Stage III or IV ovarian cancer	Randomized Phase 1/2	Experimental arm: IV ascorbate dose escalation initiated at 15 g per infusion up to 75 or 100 g with a target plasma concentration (20–23 mM); Once therapeutic dose established, ascorbate administered twice weekly with chemotherapy for 6 months, and IV ascorbate treatment alone continued for another 6 months Control arm: No IV ascorbate	Experimental arm: Carboplatin and paclitaxel treatment for 6 months Control arm: Carboplatin and paclitaxel treatment for 6 months	Patients in the ascorbate arm compared to the control chemotherapy arm had decreases in nearly all categories of adverse events including neurotoxicity, bone marrow toxicity, hepatobiliary toxicity, infection, renal toxicity, pulmonary toxicity, and gastrointestinal toxicity. Ascorbate did not increase grade 3 or 4 toxicity. Grades 1 and 2 toxicities were substantially decreased in the IV ascorbate treatment arm compared to the control arm	Median time for progression-free survival increased 8.75 months in ascorbate treatment arm compared to the control arm. Trend toward improved overall survival in the IV ascorbate treatment group, but trial not statistically powered to detect efficacy. Twenty-two out of 27 patients completed the study.

(Continued)

TABLE 7.2 (*Continued*)
High-dose and pharmacologic intravenous vitamin C clinical trials (greater than 1 g daily intravenous dose)[a]

National Clinical Trial (NCT) Number	Reference	n	Patient Diagnosis	Trial Design/ Phase	Intravenous (IV) Ascorbate (Vitamin C) Dose and Frequency	Concurrent Treatment Dose	Toxicity	Reported Outcomes/ Conclusions
NCT01050621	[137] Hoffer, 2015	14 enrolled	Advanced cancer (including lung, rectum, colon, bladder, ovary, cervix, tonsil, breast, biliary tract)	Single arm Phase 1/2	IV ascorbate: 1.5 g/kg when body mass index (BMI) was 30 kg/m^2 or less and normalized to body weight corresponding to BMI 24 kg/m^2 for patients with a BMI >30; infusions administered three times weekly (at least 1 day apart) when chemotherapy was administered (but not on the same day as IV chemotherapy) and any 2 days per week during weeks when no chemotherapy was given	Standard care chemotherapy	IV ascorbate was nontoxic for all participants and acceptably safe. Increased thirst and increased urinary flow were common minor symptoms during all ascorbate infusions. Three patients experienced unpleasant side effects of the ascorbate infusions: nausea and vomiting during the infusions (n = 1), unpleasant feeling in upper abdomen during infusions, mental haze afterward (n = 1), chills and headache during infusions, and increased leg edema following infusions	Three out of 14 patients experienced transient stable disease, and another three patients experienced longer-lasting but impermanent stable disease with unusually favorable clinical trajectories deemed highly unlikely to result from chemotherapy alone (probability less than 33%). These three cases could represent exceptional responses for future studies that focus on cervix, biliary tract, and head and neck cancer types. Six out of 14 patients experienced neither objective nor subjective benefit. The remaining two enrolled patients deteriorated rapidly in ways that could not reasonably be attributed to ascorbate therapy.

(*Continued*)

TABLE 7.2 (*Continued*)

High-dose and pharmacologic intravenous vitamin C clinical trials (greater than 1 g daily intravenous dose)[a]

National Clinical Trial (NCT) Number	Reference	n	Patient Diagnosis	Trial Design/ Phase	Intravenous (IV) Ascorbate (Vitamin C) Dose and Frequency	Concurrent Treatment Dose	Toxicity	Reported Outcomes/ Conclusions
								The six patients with colorectal cancer had disappointing clinical trajectories, suggesting that chemotherapy-resistant advanced colorectal cancer might be a poor target for future clinical trials of IV ascorbate plus conventional chemotherapy treatment.
NCT01364805	[93] Polireddy, 2017	14 enrolled (12 out of 14 entered phase 2a)	Locally advanced or metastatic pancreatic cancer	Single arm Phase 1/2	Phase 1: Escalating doses of IV ascorbate (25–125 g) one to three times weekly for 4 weeks	Gemcitabine (1000 mg/m^2 (with a few subjects receiving reduced doses as determined by the treating oncologist) once weekly for 2 consecutive weeks followed by 1 week of rest until disease progression)	Adverse events attributable to IV ascorbate were grade 1 nausea and thirst. No other adverse events were found to be related to IV ascorbate	Of the 12 participants who completed phase 2a treatment, 50% (6 of 12) survived over 1 year, and 8.3% (1 of 12) survived more than 2 years after diagnosis. Median progression-free survival was 3 months. Median overall survival was 15.1 months. Patients experienced a mix of stable disease, partial response, and disease progression: six patients (50%) had disease progression; one had tumor response and was removed from the trial for surgical resection; four dropped out due to unrelated medical issues; one dropped out for personal reasons. IV ascorbate was safe in patients and demonstrated the possibility to prolong patient survival. No clinically relevant interference with gemcitabine pharmacokinetics from IV ascorbate administration.

(Continued)

TABLE 7.2 (*Continued*)

High-dose and pharmacologic intravenous vitamin C clinical trials (greater than 1 g daily intravenous dose)[a]

National Clinical Trial (NCT) Number	Reference	n	Patient Diagnosis	Trial Design/ Phase	Intravenous (IV) Ascorbate (Vitamin C) Dose and Frequency	Concurrent Treatment Dose	Toxicity	Reported Outcomes/ Conclusions
NCT01080352	[97] Nielsen, 2017	23 enrolled	Chemotherapy-naïve, castration-resistant metastatic prostate cancer	Single arm Phase 2	IV ascorbate (5 g once during week 1, 30 g once during week 2, and 60 g once weekly during weeks 3–12) Patients may continue with weekly 60 g IV treatments in weeks 13–20 A daily oral dose of 500 mg vitamin C starting from the first infusion for 26 weeks was administered to avoid a hypothetical rebound deficiency following infusion	None	Fifty-three adverse events were recorded. Three adverse events related to fluid load were directly related to IV ascorbate. Eleven were graded as serious due to hospitalization, although three were elective procedures. The most frequent adverse events were hypertension and anemia. Two patients experienced pulmonary embolism	Twenty patients completed efficacy evaluation. No signs of disease remission were observed. No patient achieved the primary endpoint of 50% reduction in prostate-specific antigen. Instead, a median increase in prostate-specific antigen of 17 µg/L was recorded at week 12. This study does not support the use of IV ascorbate outside clinical trials.
NCT00284427	None given	25 enrolled	Ovarian, cervical, or uterine cancer	Single arm Phase 2	No dose provided IV ascorbate given two to three times a week	None	Not available	Not available

(*Continued*)

TABLE 7.2 (*Continued*)

High-dose and pharmacologic intravenous vitamin C clinical trials (greater than 1 g daily intravenous dose)[a]

National Clinical Trial (NCT) Number	Reference	n	Patient Diagnosis	Trial Design/ Phase	Intravenous (IV) Ascorbate (Vitamin C) Dose and Frequency	Concurrent Treatment Dose	Toxicity	Reported Outcomes/ Conclusions
NCT02655913	[153] Ou, 2017	97 enrolled	Stage III–IV non-small cell lung cancer	Randomized Phase 1/2	Phase 1: IV ascorbate (1 g/kg, 1.2 g/kg, or 1.5 g/kg three times a week for 8 weeks)	Modulated Electrohyperthermia: 150W × 60 min/ session, three times a week for 8 weeks administered either before, during, or after IV ascorbate treatment	Phase 1 results (n = 15): grade 3 diarrhea at 1.5 g/kg/day dose (1 out of 15 patients). Minor grade 1 fatigue and nausea experienced. No hematologic or serum creatinine abnormalities discovered during course of trial	Phase 1 results (n = 15): High-dose IV ascorbate simultaneous treatment with modulated electrohyperthermia is safe, well tolerated, and significantly increases plasma ascorbate levels for non-small cell lung cancer patients. Fasting plasma ascorbate concentration was significantly higher in the simultaneous treatment with modulated electrohyperthermia compared to ascorbate-only treatment or ascorbate treatment before or after modulated electrohyperthermia.

(Continued)

TABLE 7.2 (*Continued*)

High-dose and pharmacologic intravenous vitamin C clinical trials (greater than 1 g daily intravenous dose)[a]

National Clinical Trial (NCT) Number	Reference	n	Patient Diagnosis	Trial Design/ Phase	Intravenous (IV) Ascorbate (Vitamin C) Dose and Frequency	Concurrent Treatment Dose	Toxicity	Reported Outcomes/ Conclusions
None given	[18] Hoffer, 2008	24 enrolled	Advanced cancer or hematologic malignancy	Single arm Phase 1	Sequential cohorts infused with IV ascorbate (0.4, 0.6, 0.9, and 1.5 g/kg three times weekly, with escalation to the next dose level on completion of one 4-week treatment cycle without dose-limiting toxicity). All patients were provided with a daily multivitamin tablet (Centrum Select, Wyeth) and 400 IU D-α-tocopherol twice daily with meals, and, on noninfusion days, 500 mg ascorbic acid twice daily to avoid large shifts in plasma ascorbate concentration	None	High-dose IV ascorbate was well tolerated. Adverse events and toxicity were minimal at all dose levels: No grade 3 or higher adverse events reported. No unusual biochemical or hematologic abnormalities observed	No patient had an objective anticancer response and all eventually experienced disease progression. Two patients at the 0.6 g/kg dose received more than six cycles of ascorbate treatment with stable disease. Patients in the 0.4 g/kg cohort had a significant deterioration in physical function, although there was no deterioration in physical function among patients in the higher-dose cohorts. No changes in social, emotional, or functional quality of life measures in any cohort. The recommended phase 2 dose of 1.5 g/kg was selected for its adequate pharmacokinetic profile and clinical practicality.

(*Continued*)

TABLE 7.2 *(Continued)*

High-dose and pharmacologic intravenous vitamin C clinical trials (greater than 1 g daily intravenous dose)[a]

National Clinical Trial (NCT) Number	Reference	n	Patient Diagnosis	Trial Design/ Phase	Intravenous (IV) Ascorbate (Vitamin C) Dose and Frequency	Concurrent Treatment Dose	Toxicity	Reported Outcomes/ Conclusions
None given	[162] Riordan, 2005	24 enrolled	Late-stage terminal cancer patients (19 out of 24 had colon or rectal primary cancer)	Single arm (No phase given)	IV ascorbate (150–710 mg/kg/day for up to 8 weeks)	None	Common adverse events (generally minor) included nausea (n = 11), dry skin or mouth (n = 7), fatigue (n = 6), and edema (n = 7). Two grade 3 adverse events possibly related to ascorbate treatment: one patient with a history of renal calculi developed a kidney stone after 13 days of treatment, and another patient experienced hypokalemia after 6 weeks of treatment	One of 24 patients had stable disease and continued the treatment for 48 weeks. The other 23 patients had progressive disease during treatment with 11 of these patients completing 8 weeks of therapy. IV vitamin C was deemed relatively safe in patients without a prior history of kidney disease. Ascorbate infusions resulted in plasma ascorbate concentrations of around 1 mM. Minimal effect of IV ascorbate treatment on blood count and chemistry parameters.
NCT01754987	None given	5 enrolled	Metastatic liver cancer	Parallel assignment Phase 1/2	Phase 1: IV ascorbate (100 g three times a week for 8 weeks) Phase 2: IV ascorbate (100 g three times a week for 16 weeks)	Sorafenib (no dose provided)	Adverse events to combined ascorbate + sorafenib treatment included dehydration (n = 1), elevated creatinine (n = 2), low platelet count (n = 2), nosebleed (n = 1), and thrush (n = 1)	Trial terminated early and no data collected.

[a] Clinical trials administering a daily dose of intravenous vitamin C greater than 1 g daily were identified as "high-dose" intravenous clinical trials.

TABLE 7.3

Low-dose intravenous vitamin C trials (1 g daily intravenous dose or less)

National Clinical Trial (NCT) Number	Reference	n	Patient Diagnosis	Trial Design/ Phase	Intravenous (IV) Ascorbate (Vitamin C) Dose and Frequency	Concurrent Treatment Dose	Toxicity	Reported Outcomes/ Conclusions	Status
None given	[108] Berenson, 2007	22 enrolled	Relapsed or refractory multiple myeloma	Single arm Phase 1/2	IV ascorbate (1 g on days 1, 4, 8, and 11 of a 21-day cycle for a maximum of eight cycles)	Bortezomib (0.7, 1.0, or 1.3 mg/m^2) and arsenic trioxide (0.125 or 0.250 mg/kg)	One occurrence of grade 4 thrombocytopenia was observed in a patient receiving high-dose bortezomib (1.3 mg/m^2). One patient had asymptomatic arrhythmia and withdrew from the study. All other adverse effects were grade 1 or 2. The treatment regimen was well tolerated	Objective responses were observed in six patients (27%); two partial and four minor responses. Median progression-free survival was 5 months, and 12-month progression-free survival was 34%. Overall survival rate was 74%. Minor response observed in one of six patients receiving the lowest dose of bortezomib (0.7 mg/m^2). Five patients responded (two partial and three minor) out of 16 patients receiving higher bortezomib doses (1.0 or 1.3 mg/m^2).	Completed

(Continued)

TABLE 7.3 (*Continued*)
Low-dose intravenous vitamin C trials (1 g daily intravenous dose or less)

National Clinical Trial (NCT) Number	Reference	n	Patient Diagnosis	Trial Design/ Phase	Intravenous (IV) Ascorbate (Vitamin C) Dose and Frequency	Concurrent Treatment Dose	Toxicity	Reported Outcomes/ Conclusions	Status
NCT00661544	[102] Qazilbash, 2008	48 enrolled	Multiple myeloma	Randomized Phase 2	Arm 1 (n=16): IV ascorbate (1000 mg over 7 days) Arm 2 (n=17): IV ascorbate (1000 mg over 7 days) Arm 3 (n=15): IV ascorbate (1000 mg over 7 days)	Arm 1 (n=16): Melphalan (100 mg/m^2 for 2 days); no arsenic trioxide Arm 2 (n=17): Melphalan (100 mg/m^2 for 2 days); arsenic trioxide (0.15 mg/kg for 7 days) Arm 3 (n=15): Melphalan (100 mg/m^2 for 2 days); arsenic trioxide (0.25 mg/kg for 7 days)	No dose-limiting toxicity seen in first 100 days postautologous hematopoietic stem cell transplantation Toxicity was limited to grade 1 or 2 nausea, vomiting, and diarrhea, which was comparable in all three arms. Grade 2 cardiac toxicities including pedal edema, hypertension, and venous thromboembolism were observed in 10% of patients. Eight patients died: six due to progressive disease, one from sepsis, and one from pancreatic cancer that was unrelated to treatment	Complete responses seen in 12 of 48 patients (25%), with an overall response rate of 85%. Median progression-free survival was 25 months. No significant difference in complete response, progression-free survival, or overall survival among the three treatment arms. Addition of arsenic trioxide and IV ascorbate to high-dose melphalan is safe and well tolerated as a preparative regimen for multiple myeloma. No engraftment failure or nonrelapse mortality seen in first 100 days postautologous hematopoietic stem cell transplantation.	Completed

(Continued)

TABLE 7.3 (*Continued*)

Low-dose intravenous vitamin C trials (1 g daily intravenous dose or less)

National Clinical Trial (NCT) Number	Reference	n	Patient Diagnosis	Trial Design/ Phase	Intravenous (IV) Ascorbate (Vitamin C) Dose and Frequency	Concurrent Treatment Dose	Toxicity	Reported Outcomes/ Conclusions	Status
None given	[107] Berenson, 2006	65 enrolled	Relapsed or refractory multiple myeloma	Single arm Phase 2	Cycle 1: IV ascorbate (1 g on days 1–4 of week 1, twice weekly during weeks 2–5 and no treatment during week 6 Cycles 2–6: IV ascorbate (1 g twice weekly in weeks 1–5 and no treatment during week 6) Patients with either stable or responsive disease at end of cycle 6 continued to receive 1 g IV ascorbate once a week	Cycle 1: Melphalan (0.1 mg/kg) and arsenic trioxide (0.25 mg/kg) days 1–4 of week 1; arsenic trioxide twice weekly during weeks 2–5; no treatment during week 6 Cycles 2–6: Melphalan (0.1 mg/kg) days 1–4 of week 1; arsenic trioxide (0.25 mg/kg) twice weekly in weeks 1–5; no treatment during week 6) Patients with either stable or responsive disease at end of cycle six continued to receive 0.25 mg/kg arsenic trioxide once a week	Grade 3/4 hematologic adverse events occurred infrequently: anemia (n = 3), neutropenia (n = 1) with only one patient requiring a dose reduction in melphalan Frequent nonhematologic grade 3/4 adverse events included fever/chills (15%), pain (8%), and fatigue (6%)	Objective responses occurred in 48% of patients (31 of 65), including 2 complete, 15 partial, and 14 minor responses. Median progression-free survival was 7 months, and overall survival was 19 months. Treatment regimen was effective, well tolerated, and a new therapeutic option for patients with relapsed or refractory multiple myeloma.	Completed

(Continued)

TABLE 7.3 (*Continued*)
Low-dose intravenous vitamin C trials (1 g daily intravenous dose or less)

National Clinical Trial (NCT) Number	Reference	n	Patient Diagnosis	Trial Design/ Phase	Intravenous (IV) Ascorbate (Vitamin C) Dose and Frequency	Concurrent Treatment Dose	Toxicity	Reported Outcomes/ Conclusions	Status
None given	[109] Abou-Jawde, 2006	20 enrolled	Relapsed or refractory multiple myeloma	Single arm Phase 2	Week 1: IV ascorbate (1000 mg on 5 consecutive days) Weeks 2–12: IV ascorbate (1000 mg twice a week) Treatment period followed by 2 weeks of rest	Week 1: Arsenic trioxide (0.25 mg/kg on 5 consecutive days); dexamethasone (40 mg on days 1 through 4) Weeks 2–12: Arsenic trioxide twice a week; dexamethasone on days 1–4 and 11–14 of each 4-week period Treatment period followed by 2 weeks of rest 14-week cycle repeated up to three cycles with dexamethasone given only on days 1–4 of each month	Treatment regimen was well tolerated with most adverse events mild or moderate. There were grade 3 adverse events in nine patients, most nonhematologic. Incidence of each grade 1 and grade 2 adverse events of fatigue, edema, nausea, hyperglycemia, and sensory neuropathy was >10%. Incidence of each grade 1 and grade 2 adverse events of anemia, thrombocytopenia, rash, arthralgias, tearing, headache, and dyspnea was <10%	Overall response rate was 30%; 80% of patients had stable disease or better. Two patients had near complete response, 4 had partial response, 10 had stable disease, and 4 had progressive disease. Median progression-free survival was 316 days in all patients and 584 days in patients with a response. Clinical efficacy of treatment regimen demonstrated.	Completed

(*Continued*)

TABLE 7.3 *(Continued)*

Low-dose intravenous vitamin C trials (1 g daily intravenous dose or less)

National Clinical Trial (NCT) Number	Reference	n	Patient Diagnosis	Trial Design/ Phase	Intravenous (IV) Ascorbate (Vitamin C) Dose and Frequency	Concurrent Treatment Dose	Toxicity	Reported Outcomes/ Conclusions	Status
None given	[110] Chang, 2009	17 enrolled	Relapsed and refractory lymphoid malignancies (12 of 17 patients had non-Hodgkin lymphoma)	Single arm Phase 2	IV ascorbate (1000 mg) for 5 days during week 1 followed by twice weekly during weeks 2–6. Cycles repeated every 8 weeks	Arsenic trioxide (0.25 mg/kg for 5 days during week 1 followed by twice weekly during weeks 2–6. Cycles repeated every 8 weeks)	Hematologic toxicities formed the majority of grade 3 and 4 toxicities. One cardiac death 3 days into treatment in a patient with known coronary artery disease. One patient with recurrent grade 4 hyperglycemia was required to discontinue treatment. Nonhematologic toxicities were generally mild, with no electrolytic imbalances or changes in renal function reported as higher than grade 2 in severity. One patient with grade 3 stomatitis	Overall median survival was 7.6 months ($n = 16$ evaluable patients). One patient with mantle cell lymphoma achieved an unconfirmed complete response after five cycles of therapy. Eight out of 16 evaluable patients did not complete cycle 1 (five were due to progressive disease, and three were due to toxicity). Among the remaining eight patients who completed at least one cycle of treatment, six had evidence of progressive disease. Arsenic trioxide and IV ascorbate treatment was generally well tolerated but had limited activity in patients with relapsed and refractory lymphoid malignancies. Trial closed early at first stage of study due to lack of activity.	Completed (trial closed early after interim analysis due to lack of activity)

(Continued)

TABLE 7.3 (*Continued*)

Low-dose intravenous vitamin C trials (1 g daily intravenous dose or less)

National Clinical Trial (NCT) Number	Reference	n	Patient Diagnosis	Trial Design/ Phase	Intravenous (IV) Ascorbate (Vitamin C) Dose and Frequency	Concurrent Treatment Dose	Toxicity	Reported Outcomes/ Conclusions	Status
None given	[111] Bael, 2008	11 enrolled	Metastatic melanoma	Single arm Phase 2	IV ascorbate (1000 mg for 5 days during week 0, and then twice weekly for an 8-week cycle)	Temozolomide (200 mg/m^2 for 5 days during weeks 1 and 5 of each 8-week cycle; and arsenic trioxide (0.25 mg/kg/day for 5 days during week 0 then twice weekly at 0.35 mg/kg for an 8-week cycle)	Common grade 1 and 2 adverse events included nausea and vomiting (n=10), fatigue (n=6), edema (n=6), rash (n=6), and elevated aspartate (AST) or alanine aminotransferase (ALT) (n=6). Grades 3 and 4 adverse events included nausea and vomiting (n=3), elevated AST or ALT (n=2), seizure (n=1), and renal failure (n = 1)	No responses seen in the first 10 evaluable patients leading to early closure of study. All patients had progressive disease with only seven patients completing a full cycle of treatment.	Completed (trial closed early after interim analysis due to lack of activity)
None given	[112] Subbarayan, 2007	5 enrolled	Refractory metastatic colorectal cancer	Single arm Phase 2	IV ascorbate (1000 mg/day for 5 days a week for 5 weeks followed by 2 weeks of rest)	Arsenic trioxide (0.25 mg/kg/day for 5 days a week for 5 weeks followed by 2 weeks of rest)	All patients developed moderate to severe toxic side effects to treatment. Grade 3 toxicities were nausea/vomiting, diarrhea, dehydration, neuropathy, thrombocytopenia, and anemia	No complete or partial remission observed (n=5). Study discontinued due to all patients developing moderate to severe toxic side effects with no measurable activity. Three patients completed at least one cycle, and two patients received more than two cycles of treatment.	Completed (trial closed early due to toxicity)

(Continued)

TABLE 7.3 (*Continued*)
Low-dose intravenous vitamin C trials (1 g daily intravenous dose or less)

National Clinical Trial (NCT) Number	Reference	n	Patient Diagnosis	Trial Design/ Phase	Intravenous (IV) Ascorbate (Vitamin C) Dose and Frequency	Concurrent Treatment Dose	Toxicity	Reported Outcomes/ Conclusions	Status
None given	[113] Wu, 2006	20 enrolled	Relapsed or refractory multiple myeloma	Single arm Phase 2	IV ascorbate (1000 mg per day for 5 days, then twice weekly during weeks 2–3)	Arsenic trioxide (0.25 mg/kg per day for 5 days, then 0.25 mg/kg twice weekly for weeks 2–3); and dexamethasone (40 mg per day orally for 5 days, then 20 mg orally twice weekly for weeks 2–3)	Grades 3–4 adverse events included bacterial infections, neutropenia, hepatic toxicity, and thrombocytopenia. Most common adverse events were bacterial infections (n = 10), peripheral edema (n = 8), fatigue (n = 7), dyspnea (n = 6), reactivation of herpes zoster (n = 5), neuropathy (n = 5), neutropenia (n = 4), thrombocytopenia (n = 3), and malaise (n = 3)	Clinical response observed in 8 of 20 patients (40%) including two partial and six minor responses. Median progression-free survival was 4 months, and median overall survival was 11 months. Fourteen out of 20 patients completed two or more cycles of treatment. This combination treatment is feasible but has moderate efficacy and significant toxicity in heavily pretreated patients with advanced multiple myeloma.	Completed

(Continued)

TABLE 7.3 (*Continued*)
Low-dose intravenous vitamin C trials (1 g daily intravenous dose or less)

National Clinical Trial (NCT) Number	Reference	n	Patient Diagnosis	Trial Design/ Phase	Intravenous (IV) Ascorbate (Vitamin C) Dose and Frequency	Concurrent Treatment Dose	Toxicity	Reported Outcomes/ Conclusions	Status
NCT00006021	[103] Bahlis, 2002	22 enrolled	Relapsed or refractory multiple myeloma	Single arm Phase 1/2	IV ascorbate (1000 mg/day for 5 days for 5 weeks followed by 2 weeks of rest) Patients allowed to receive a maximum of six treatment cycles	Phase 1 treatment groups: Arsenic trioxide (0.15 mg/kg/day or 0.25 mg/kg/day for 5 days for 5 weeks followed by 2 weeks of rest) Phase 2 treatment: Arsenic trioxide (0.25 mg/kg/day for 5 days for 5 weeks followed by 2 weeks of rest)	Phase 1 toxicity: One episode of grade 3 hematologic toxicity (leukopenia) was observed. No grade 3 nonhematologic toxicities or thrombocytopenia were observed. Phase 1 and 2 patients had a similar toxicity profile	Phase 1 results (n = 6): Two patients had partial responses; four had stable disease. Treatment regimen had acceptable toxicity with promising evidence of activity in refractory or relapsed myeloma. Ascorbate coadministration did not alter arsenic pharmacokinetics or increased toxicity, although it clearly modulated intracellular glutathione levels.	Completed

(Continued)

TABLE 7.3 *(Continued)*

Low-dose intravenous vitamin C trials (1 g daily intravenous dose or less)

National Clinical Trial (NCT) Number	Reference	n	Patient Diagnosis	Trial Design/ Phase	Intravenous (IV) Ascorbate (Vitamin C) Dose and Frequency	Concurrent Treatment Dose	Toxicity	Reported Outcomes/ Conclusions	Status
None given	[114] Held, 2013	10 enrolled	Relapsed/ refractory multiple myeloma	Single arm Phase 1/2	IV ascorbate (1 g daily for 3 days starting day 7 of cycle 1 only, then twice weekly for 2 weeks) Responders could continue treatment for up to eight 3-week cycles	Arsenic trioxide (0.25 mg/kg for 3 days starting on day 7 of cycle 1 only, then once a week for 2 weeks); and bortezomib (escalating IV dose of 1 mg/m^2 or 1.3 mg/m^2 on days 1 and 8 of a 21-day cycle)	No dose-limiting toxicities. Most toxicities were grades 1 and 2 including fatigue, nausea, constipation, anorexia, and peripheral neuropathy. There were three grades 3 and 4 adverse events: grade 3 neutropenia ($n = 1$), grade 4 thrombocytopenia ($n = 1$), and grade 3 transaminitis ($n = 1$)	Four out of 10 patients (40%) had clinical benefit lasting approximately 6 months: three patients with stable disease and one patient achieving a durable partial response. Study was stopped prematurely after the completion of cohort 2 due to poor accrual.	Completed (trial stopped early due to poor accrual)

(Continued)

TABLE 7.3 (*Continued*)
Low-dose intravenous vitamin C trials (1 g daily intravenous dose or less)

National Clinical Trial (NCT) Number	Reference	n	Patient Diagnosis	Trial Design/ Phase	Intravenous (IV) Ascorbate (Vitamin C) Dose and Frequency	Concurrent Treatment Dose	Toxicity	Reported Outcomes/ Conclusions	Status
NCT00671697	[104] Welch, 2011	13 enrolled	Myelodysplastic syndrome or acute myeloid leukemia	Single arm Phase 1	IV ascorbate (1000 mg for 5 days followed by weekly doses for 15 weeks)	Decitabine (20 mg/m^2 daily on days 1–5 every 28 days for four 28-day cycles) and arsenic trioxide escalated in three dose cohorts (0.1 mg/kg for 5 days followed by weekly doses for 15 weeks; 0.2 mg/kg for 5 days followed by weekly doses for 15 weeks; or 0.3 mg/kg for 5 days followed by weekly doses for 15 weeks)	Grade 3 QTc prolongation in a patient of the second cohort (0.2 mg/kg arsenic trioxide treatment) was a dose-limiting toxicity. Two other dose-limiting toxicities occurred in patients in the third cohort (0.3 mg/kg arsenic trioxide dose) who had prior neutropenia before enrollment in the study and developed sepsis with hypotension and hypoxia and died. In total, four patients discontinued therapy due to toxicities	One patient had morphologic complete remission with incomplete blood count recovery. Five patients had stable disease after four cycles of treatment. Four patients had progressive disease. Three patients did not complete cycle 2 and were not evaluated for disease response. 0.2 mg/kg identified as maximum tolerated dose of arsenic trioxide in combination with decitabine and ascorbate.	Completed

(Continued)

TABLE 7.3 (*Continued*)
Low-dose intravenous vitamin C trials (1 g daily intravenous dose or less)

National Clinical Trial (NCT) Number	Reference	n	Patient Diagnosis	Trial Design/ Phase	Intravenous (IV) Ascorbate (Vitamin C) Dose and Frequency	Concurrent Treatment Dose	Toxicity	Reported Outcomes/ Conclusions	Status
NCT00621023	None given	7 enrolled	Myelodysplastic syndrome	Single arm Phase 2	IV ascorbate (1000 mg on days 1–5 in cycle 1, then twice weekly for remaining cycles)	Decitabine (20 mg/m^2 on days 1–5 of each cycle) and arsenic trioxide (0.25 mg/kg on days 1–5 in cycle 1 followed by 0.25 mg/kg twice weekly for remaining cycles)	Four out of six enrolled patients had unacceptable toxicity. Serious adverse events included febrile neutropenia ($n = 2$), infection ($n = 6$), cardiac pain ($n = 1$), hematoma ($n = 1$), and pleural effusion ($n = 1$)	Five out of six enrolled patients died, and follow-up was discontinued for the remaining patient. No patients responded.	Completed (enrollment stopped early)

(*Continued*)

TABLE 7.3 (*Continued*)

Low-dose intravenous vitamin C trials (1 g daily intravenous dose or less)

National Clinical Trial (NCT) Number	Reference	n	Patient Diagnosis	Trial Design/ Phase	Intravenous (IV) Ascorbate (Vitamin C) Dose and Frequency	Concurrent Treatment Dose	Toxicity	Reported Outcomes/ Conclusions	Status
NCT00469209	[105] Sharma, 2012	60 enrolled	Multiple myeloma	Randomized Phase 1/2	Arm 1: IV ascorbate (1000 mg daily IV for 7 days) Arm 2: IV ascorbate (1000 mg daily for 7 days) Arm 3: IV ascorbate (1000 mg daily IV for 7 days)	Arm 1: No bortezomib; melphalan (100 mg/m^2); arsenic trioxide (0.25 mg/kg for 7 days) Arm 2: bortezomib (1.0 mg/m^2); melphalan (100 mg/m^2); arsenic trioxide (0.25 mg/kg for 7 days) Arm 3: bortezomib (1.5 mg/m^2); melphalan (100 mg/m^2); arsenic trioxide (0.25 mg/kg for 7 days)	The most common adverse events were fluid retention (78%), nausea (65%), and diarrhea (63%). Grade 3–4 toxicities were similar in all three treatment arms, with 12 patients (60%) in arm three experiencing left ventricular dysfunction (n=1), chest pain (n=1), hypertension (n=1), pulmonary edema (n=2), pleural effusion (n=1), dyspnea (n=1), mucositis (n=1), intestinal obstruction (n=1), and elevated transaminases (n=1)	Complete response rates in arms 1, 2, and 3 were 20%, 10%, and 10%, respectively. Median progression-free survival in arms 1, 2, and 3 was 17.8, 17.4. and 20.7 months, respectively. Progression-free survival and overall survival were significantly shorter in patients with high-risk cytogenetics and relapsed disease in all treatment groups. Adding bortezomib to a regimen of arsenic trioxide, ascorbic acid, and high-dose melphalan is safe and well tolerated in patients with multiple myeloma with no increase in grade 3–4 toxicity or treatment-related mortality. There was no significant improvement in complete response rate, progression-free survival, or overall survival in arms 2 and 3 with bortezomib treatment.	Completed

(Continued)

TABLE 7.3 (*Continued*)

Low-dose intravenous vitamin C trials (1 g daily intravenous dose or less)

National Clinical Trial (NCT) Number	Reference	n	Patient Diagnosis	Trial Design/ Phase	Intravenous (IV) Ascorbate (Vitamin C) Dose and Frequency	Concurrent Treatment Dose	Toxicity	Reported Outcomes/ Conclusions	Status
NCT00258245	None given	5 enrolled	Relapsed or refractory multiple myeloma or plasma cell leukemia	Single arm Phase 1	IV ascorbate (1000 mg on days 1, 4, 8, and 11)	Thalidomide (50 mg/day); dexamethasone (40 mg on days 1, 4, 8, and 11); Escalating doses of bortezomib (0.7 mg/m^2 or 1.0 mg/m^2 on days 1, 4, 8, and 11); escalating doses of arsenic trioxide (0.10 mg/kg, 0.15 mg/kg, or 0.25 mg/kg on days 1, 4, 8, and 11); aspirin (325 mg daily)	Not available	Not available	Completed

TABLE 7.4

Oral vitamin C administration clinical trials (see https://sysrev.com/p/7162)

National Clinical Trial (NCT) Number	Reference	n	Patient Diagnosis	Trial Design/ Phase	Oral Ascorbate (Vitamin C) Dose and Frequency	Concurrent Treatment Dose	Toxicity	Reported Outcomes/ Conclusions	Status
NCT03397173	None given	28 anticipated	Myelodysplastic syndromes, myeloproliferative neoplasm, and acute myeloid leukemia	Single arm Phase 2	Oral ascorbate (1 g/day 3 days prior to start of azacitidine treatment then continues daily for a total of 28 days)	Azacitidine (75 $mg/m^2/day$ for 7 consecutive days)	Not available	Not available	Recruiting
NCT03682029	None given	70 anticipated	Myelodysplastic syndromes, chronic myelomonocytic leukemia, and cytopenia	Randomized (no study phase provided)	Experimental arm: Oral ascorbate (1000 mg/daily for 12 months; dose separated into two 500 mg/capsules daily)	None	Not available	Not available	Recruiting
					Placebo arm: two oral placebo capsules daily for 12 months	None			

(Continued)

TABLE 7.4 *(Continued)*

Oral vitamin C administration clinical trials (see https://sysrev.com/p/7162)

National Clinical Trial (NCT) Number	Reference	n	Patient Diagnosis	Trial Design/ Phase	Oral Ascorbate (Vitamin C) Dose and Frequency	Concurrent Treatment Dose	Toxicity	Reported Outcomes/ Conclusions	Status
NCT03624270	None given	60 anticipated	Acute promyelocytic leukemia	Single arm Phase 2	Induction phase: Oral ascorbate (1 g daily for 42 days) Consolidation phase: Oral ascorbate (1 g daily for 14 days every 28 days for two cycles) Maintenance phase (for all patients): Oral ascorbate (1 g daily for 2 weeks every 2 months for 24 months)	Induction phase: Arsenic trioxide (10 mg daily for 42 days); all-trans retinoic acid (45 mg/m^2 daily in divided doses for 42 days); daunorubicin (50 mg/m^2 daily for 3 days) Consolidation phase: Arsenic trioxide (10 mg daily for 14 days every 28 days for two cycles); all-trans retinoic acid (45 mg/m^2 daily for 14 days every 28 days for two cycles); daunorubicin (50 mg/m^2 daily for 2 days every 4–6 weeks for two cycles); cytarabine (100 mg/m^2 daily for 5 days every 4–6 weeks for two cycles) Maintenance phase (for all patients): Arsenic trioxide (10 mg daily for 2 weeks every 2 months for 24 months); all-trans retinoic acid (45 mg/m^2 daily for 2 weeks every 2 months for 24 months)	Not available	Not available	Recruiting

(Continued)

TABLE 7.4 *(Continued)*

Oral vitamin C administration clinical trials (see https://sysrev.com/p/7162)

National Clinical Trial (NCT) Number	Reference	n	Patient Diagnosis	Trial Design/ Phase	Oral Ascorbate (Vitamin C) Dose and Frequency	Concurrent Treatment Dose	Toxicity	Reported Outcomes/ Conclusions	Status
NCT01891747	None given	12 enrolled	Malignant glioma	Single arm Phase 1	Oral ascorbate (250 mg tablet once daily)	L-methylfolate (15, 30, 60, or 90 mg daily); bevacizumab (10 mg/kg every 14 days); temozolomide (150 mg/m^2 daily for 5 days per month)	Not available	Not available	Active, not recruiting
NCT00317811	[118] Berenson, 2009	35 enrolled	Newly diagnosed multiple myeloma	Single arm Phase 2	Oral ascorbate (1 g on days 1–4 of 28-day cycle)	Bortezomib (1.0 mg/m^2 on days 1, 4, 8, and 11 followed by a 17-day rest period; maintenance treatment of bortezomib (1.3 mg/m^2 every 2 weeks until progression); melphalan (0.1 mg/kg on days 1–4 of 28-day cycle)	Grades 3 and 4 adverse events occurred in 17 and 5 patients, respectively. Grades 3 and 4 adverse events considered drug related with occurrences in more than one patient were neutropenia (n = 3), peripheral neuropathy (n = 3), and thrombocytopenia (n = 3). Drug-related serious adverse events included grade 3 thrombocytopenia (n = 1), grade 2 atrial flutter (n = 1), and grade 2 fatigue (n = 1).	Responses occurred in 23 of 31 evaluable patients (74%) including 5 (16%) complete, 3 (10%) very good partial, 6 (19%) partial, and 9 (29%) minimal responses. Six patients (19%) had stable disease. Disease control was achieved in 29 (94%) patients. Median time to progression was 19 months. This combination treatment is an efficacious and well-tolerated frontline treatment for multiple myeloma patients.	Completed

(Continued)

TABLE 7.4 *(Continued)*

Oral vitamin C administration clinical trials (see https://sysrev.com/p/7162)

National Clinical Trial (NCT) Number	Reference	n	Patient Diagnosis	Trial Design/ Phase	Oral Ascorbate (Vitamin C) Dose and Frequency	Concurrent Treatment Dose	Toxicity	Reported Outcomes/ Conclusions	Status
NCT00274820	[106] Bejanyan, 2012	28 enrolled	Myelodysplastic/ myeloproliferative neoplasms, primary myelofibrosis, and chronic myelomonocytic leukemia	Single arm Phase 2	Oral ascorbate (1000 mg for 5 days in first week followed by twice weekly for 11 weeks)	Arsenic trioxide (0.25 mg/kg for 5 days in first week followed by twice weekly for 11 weeks); dexamethasone (4 mg daily for 5 days every 4 weeks); thalidomide (50 mg daily for 2 weeks; then 100 mg daily up to 6 months, if tolerated)	Most adverse events were mild (grade 1) to moderate (grade 2). Seven out of 28 enrolled patients (25%) experienced grade ≥3 hematologic toxicities, including thrombocytopenia (n = 3), neutropenia (n = 3), and leukocytosis (n = 1). Nonhematologic grades 3 and 4 adverse events included dyspnea (18%) and infections (14%). Fatigue was the most common nonhematologic toxicity observed in 82% of patients. About two-thirds of patients developed neuropathy (most mild to moderate) that all resolved. No thrombotic episodes	Six out of 28 patients (21%) responded to treatment (seen only in patients who completed the full 12-week cycle of therapy): 1 patient achieved partial response and 5 patients had clinical improvement. All five patients with *JAK2V617F* mutation had stable disease throughout the study period. One out of two patients with a *TET2* mutation achieved hematologic improvement, and the other patient had stable disease. Study treatment regimen was relatively well tolerated and yielded clinical responses in patients with primary myelofibrosis and myelodysplastic/ myeloproliferative neoplasms.	Completed

(*Continued*)

TABLE 7.4 *(Continued)*

Oral vitamin C administration clinical trials (see https://sysrev.com/p/7162)

National Clinical Trial (NCT) Number	Reference	n	Patient Diagnosis	Trial Design/ Phase	Oral Ascorbate (Vitamin C) Dose and Frequency	Concurrent Treatment Dose	Toxicity	Reported Outcomes/ Conclusions	Status
NCT02877277	[119] Klein, 2017	20 enrolled	Myelodysplastic syndrome and acute myeloid leukemia	Randomized Pilot study	Experimental arm: Oral ascorbate (500 mg daily for 56 days)	Azacitidine (no dose provided)	Not available	Not available	Completed
					Placebo arm: Oral placebo tablet for 56 days	Azacitidine (no dose provided)			

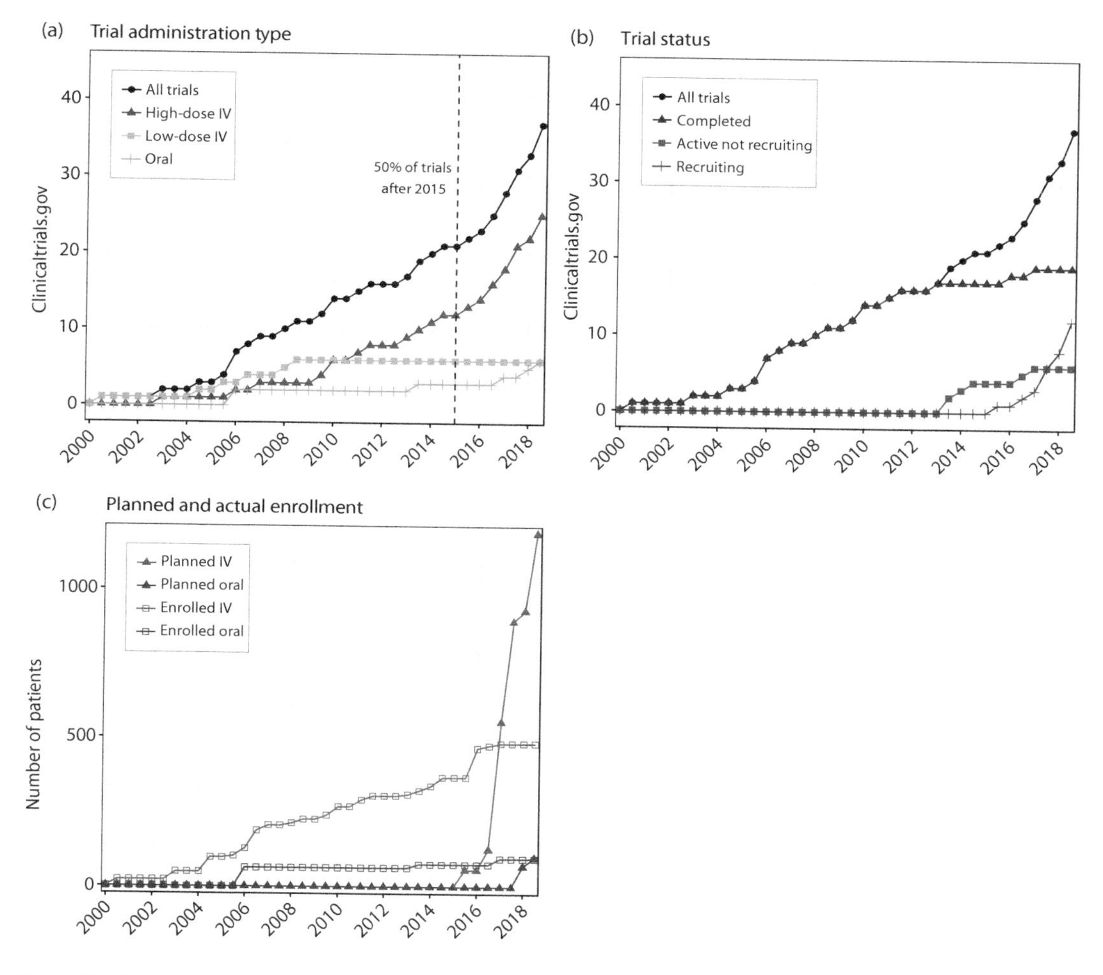

Figure 7.1. Number of vitamin C cancer clinical trials in ClinicalTrials.gov as of November 29, 2018, by (a) vitamin C administration (intravenous [IV] or oral), (b) trial status, and (c) planned or actual patient enrollment. (a) The black top line includes all cancer clinical trials that used vitamin C as a treatment intervention (either IV or oral administration). "Low-dose IV" was categorized as IV ascorbate dosing at 1 g daily and below. "High-dose IV" included any IV ascorbate dose above 1 g daily. (b) "Completed" trials describe studies that have ended with participants no longer being treated. "Active, not recruiting" trials are ongoing, but new participants are not being enrolled. "Recruiting" trials are actively recruiting participants. (c) Patients in clinical trials categorized as "complete" in the ClinicalTrials.gov database were identified as "enrolled." Patient enrollment planned for recruiting trials are marked by the "planned" lines. Note that patients enrolled in actively accruing trials were not captured in this figure.

Intravenous ascorbate dosing in cancer clinical trials analyzed ranged from 1 g daily to approximately 100 g three times a week. Evidence suggests that the preferable pharmacologic ascorbate dose is 1 g/kg administered intravenously for 2 hours, at a minimum frequency of twice weekly [19,49,66,99–102], and the suggested minimum pharmacologic ascorbate treatment duration before assessing efficacy is at least 2 months (3–4 months is preferable for minimum duration) [49,99,103–105]. There is evidence that doses of pharmacologic ascorbate near 1 g/kg (i.e., often >50 g/day) are needed for anticancer activity, while lower doses (≤10 g/day) perhaps improve quality of life [95]. In animal experiments, multiple infusions of ascorbate also appear to be more effective compared to a single infusion of ascorbate [43,106]. Despite limits of translating mice data to humans, daily intraperitoneal injections of ascorbate in mice, where intraperitoneal administration achieves pharmacologic concentrations [70], hindered tumor growth far more than when injections were performed every other day [106]. In humans, it is unclear whether continuous infusion is comparable to 2-hour infusions several times weekly [107,108].

There were 1390 planned patients and 491 enrolled patients in clinical trials using intravenously administered vitamin C as a treatment as of November 29, 2018 (Figure 7.1c). Clinical trials using orally administered vitamin C as a treatment had 158 anticipated patients and enrolled 95 patients (Table 7.4). The dramatic increase in planned intravenous trial patients in the past year can be attributed to both an increase in the number of intravenous trials and the number of planned patients per trial. It is unclear how often sponsors and principal investigators update ClinicalTrials.gov with accrual information, so it is difficult to determine whether currently active trials have issues with timely accrual. In the 2000–2016 period, there were only 8 terminated trials compared to 19 completed trials, suggesting that cancer vitamin C trials have been able to enroll adequate numbers of participants. When trials are terminated for low accrual, one possible cause may be patient reluctance to spend multiple hours on multiple days each week undergoing infusion.

There were nine trials with vitamin C (ascorbic acid) as the only investigational agent, with five of these trials open and actively enrolling participants (Figure 7.2a and b). Thirty-three trials combined standard therapy agents with vitamin C (either intravenous or oral administration) as an adjunct treatment. Among these 33 trials, 7 trials used three or more agents in combination with vitamin C; 11 trials used two agents in combination with vitamin C; and 15 trials used only one other agent in combination with vitamin C (Figure 7.3). The most common agents combined with vitamin C are arsenic trioxide and gemcitabine (eight trials each). A total of 183 participants were enrolled across all seven completed arsenic trioxide trials (six trials were used in combination with low-dose intravenous ascorbate, and one trial was used in combination with oral ascorbate) [109–113]. An additional eight trials without an identifying NCT number treated patients with arsenic trioxide and ascorbate (Table 7.2) [114–121]. There is one actively recruiting arsenic trioxide trial (NCT03624270) with 60 anticipated patients. The combination of vitamin C and arsenic trioxide is frequently administered in refractory multiple myeloma or leukemia to enhance the efficacy and tolerability of the chemotherapeutic agent [122]. Mechanistically, vitamin C is shown to enhance cytotoxicity in multiple myeloma cells from arsenic trioxide by decreasing intracellular glutathione [123]. The 1 g daily dose of ascorbate used in many arsenic trioxide trials is not pharmacologic. The original justification of the chosen dose was to produce tissue saturation, based on oral dosing data [124]. While such low-dose intravenous vitamin C may improve quality of life and even result in extended chemotherapy tolerance, ascorbate in this setting should not be considered to have direct anticancer activity [92]. Any observed anticancer activity should instead be attributed to arsenic trioxide and not to low-dose vitamin C in these trials [124]. A more comprehensive list of low-dose intravenous vitamin C trials is provided in Table 7.3.

Pancreatic cancer and leukemia have been the most prevalent diseases studied in vitamin C cancer clinical trials (nine trials of each disease), with many trials still actively recruiting participants. While all nine pancreatic cancer trials administered pharmacologic ascorbate, most of the leukemia trials used oral vitamin C. Myelodysplastic syndromes were the next most prevalent disease category studied in vitamin C trials (eight trials: four completed and four recruiting), with half using oral vitamin C. Disease types investigated in oral ascorbate trials were leukemia, multiple myeloma, glioma, and myelodysplastic syndrome [113,125,126]. Most patients in cancer clinical trials had advanced/metastatic disease (26 out of 42 trials) where often the aim is to increase the duration of treatment and life span of patients undergoing toxic treatments through complementary vitamin C treatment. Actively recruiting trials have a greater number of anticipated participants compared to the number of enrolled participants from completed intravenous trials. Out of 17 actively recruiting trials, most (11 trials) have an anticipated enrollment of over 50 participants. In contrast, out of 16 completed intravenous trials, most (10 trials) enrolled 25 or fewer participants.

CLINICAL EFFICACY AND SAFETY OF INTRAVENOUS ASCORBATE IN CANCER PATIENTS

The pharmacokinetic differences between oral and pharmacologic intravenous dosing initiated an interest in developing clinical studies to

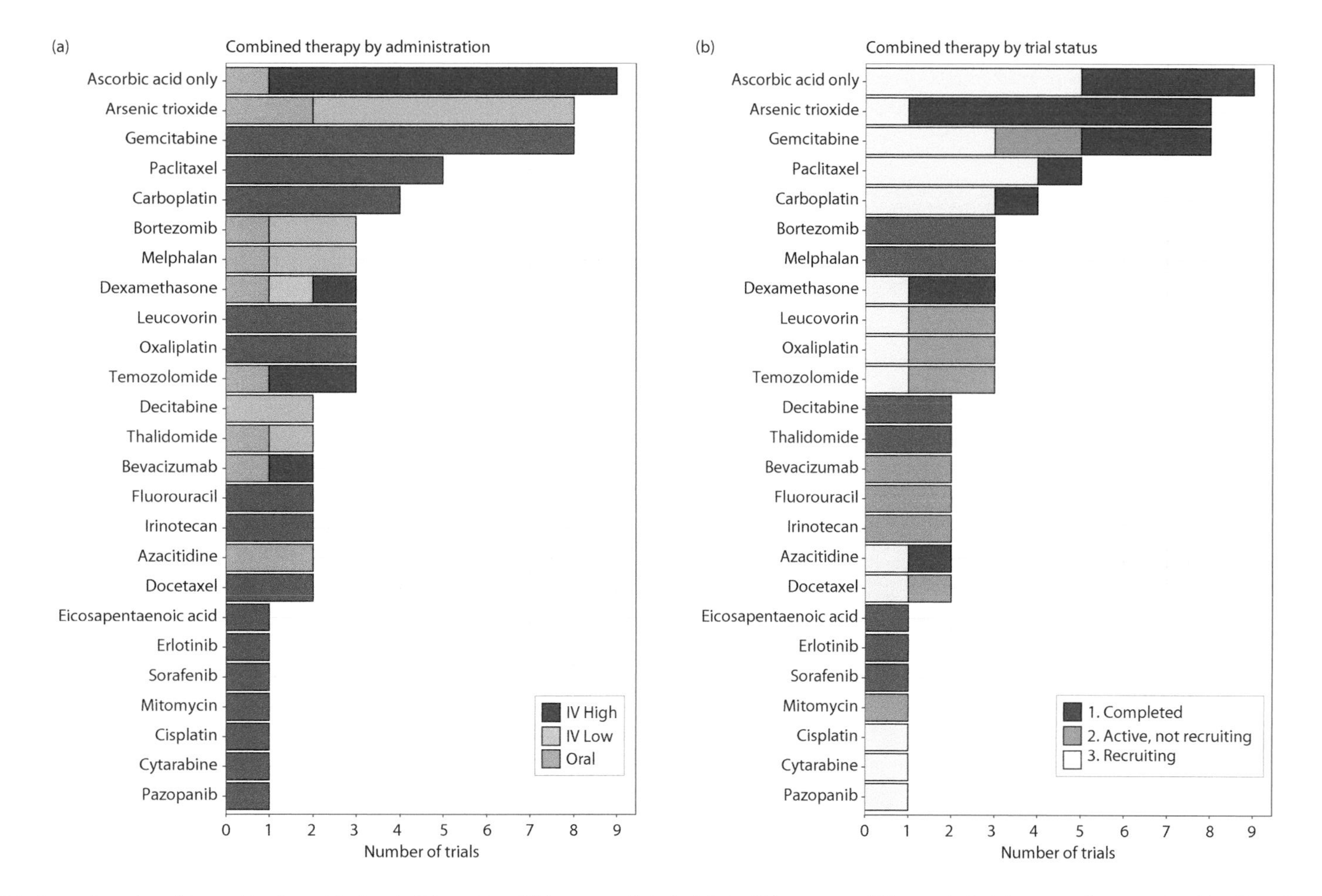

Figure 7.2. Agents investigated in combination with vitamin C (ascorbic acid) in cancer clinical trials by (a) intravenous [IV] or oral administration of vitamin C and by (b) trial status as of November 29, 2018. "IV Low" dosing was categorized as IV ascorbate dosing at 1 g daily and below. "IV High" included any IV ascorbate dose above 1 g daily. "Completed" trials have ended with participants no longer being treated. "Active, not recruiting" trials are ongoing, but new participants are not being enrolled. "Recruiting" trials are actively recruiting participants. The agents used in combination with vitamin C in cancer clinical trials are organized in the bar charts from top to bottom by prevalence. Note that the same trial can have more than one investigational agent and would be represented more than once in this figure under each agent used in the trial.

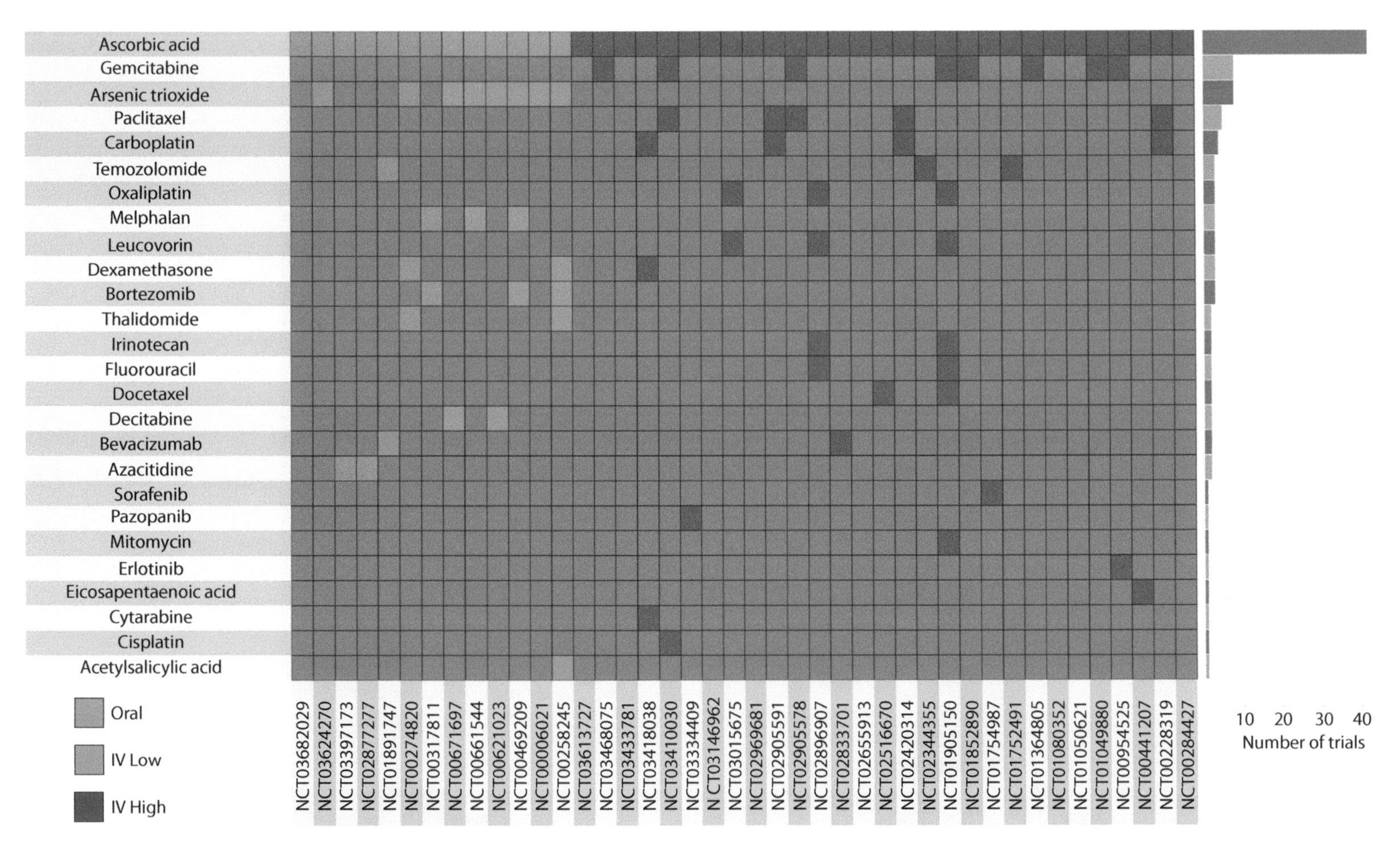

Figure 7.3. Agents investigated in combination with vitamin C (ascorbic acid) in cancer clinical trials with trials listed by ClinicalTrials.gov identifier (NCT number). Trials are color-coded for the type of vitamin C administration used (i.e., high-dose intravenous [IV], low-dose IV, or oral). "IV Low" dosing was categorized as IV ascorbate dosing at 1 g daily and below. "IV High" included any IV ascorbate dose above 1 g daily. Normalized drug names were extracted from drugbank.ca and imported via https://sysrev.com/p/6963. Only clinical trials extracted from ClinicalTrials.gov with affiliated NCT numbers were used in this figure.

investigate the ability to reach pharmacologic levels of ascorbate in humans easily and safely and to determine any potential therapeutic benefit. An increasing number of phase 1 and 2 clinical trials have been initiated to test the safety and efficacy of pharmacologic ascorbate, mostly as an adjunct treatment for various cancers (Table 7.2). In a phase 1 trial of patients with metastatic pancreatic cancer treated with pharmacologic ascorbate, gemcitabine, and erlotinib, eight out of nine patients had tumor shrinkage after 8 weeks of treatment with no increased toxicity from the addition of pharmacologic ascorbate treatment [99]. Pharmacologic ascorbate reduced nearly all categories of toxicities, including neurotoxicity, bone marrow toxicity, hepatobiliary toxicity, infection, renal toxicity, pulmonary toxicity, and gastrointestinal toxicity in a randomized trial of advanced ovarian cancer patients treated with carboplatin and paclitaxel chemotherapy [66]. Even at lower intravenous ascorbate treatment doses (1 g on days 1, 4, 8, and 11 of a 21-day cycle) in combination with bortezomib and arsenic trioxide, a phase 1/2 trial of relapsed or refractory multiple myeloma patients reported objective responses in 6 out of 22 patients (27%): 2 patients with partial responses and 4 patients with minor responses [115]. More recently, a phase 1 study of 1.5 g/kg/day for 3 days in a 14-day cycle of mFOLFOX6 and FOLFIRI in 36 patients with metastatic colorectal cancer reported a provocative 58% objective response rate and a 96% disease control rate with no difference between patients with wild-type and mutant RAS and BRAF [127].

According to data extracted from ClinicalTrials.gov, there were 35 intravenous ascorbate clinical trials: 29 trials used pharmacologic doses, while 6 used low doses of intravenous ascorbate (Figure 7.4). The completed phase 1 trials that used pharmacologic ascorbate reported no adverse events beyond those from the disease or from concomitant chemotherapy treatment [19,99,102,128]. In one trial, pharmacologic ascorbate significantly decreased grade 1 and 2 chemotherapy-related toxicities from carboplatin and paclitaxel in patients with stage III–IV ovarian cancer [66]. For cost reasons, retrospective controls were often used in safety trials [2], making data interpretation difficult. In one phase 1 trial, adding pharmacologic ascorbate daily to radiation therapy was safe for locally advanced pancreatic cancer, and while the study was not powered to indicate efficacy, progression-free survival and overall survival increased compared to similar patients in a retrospective comparison [129]. Using a retrospective control group is not ideal due to variation in chemotherapy regimens and differences in initial carbohydrate antigen 19-9 levels [129]. Despite issues with the use of retrospective controls, intravenous vitamin C has proven to be remarkably safe, although a few circumstances call for caution. Two contraindications for pharmacologic ascorbate treatment in patients have been identified. First, since ascorbate metabolites are only excreted from the body through urination, pharmacologic ascorbate treatment is contraindicated for patients with compromised renal function [130], due to the possibilities of worsening kidney function and/or precipitating oxalate renal stone formation from an inability of the kidneys to excrete oxalate, an end product of ascorbate catabolism [131–133]. Second, hemolytic anemia has been reported in patients who are deficient in glucose-6-phosphate dehydrogenase who received pharmacologic ascorbate treatment [68,134]. Patients should be screened for glucose 6 phosphate dehydrogenase deficiency prior to treatment to avoid clinical complications. A history of paroxysmal nocturnal hemoglobinuria is also a contraindication for intravenous ascorbate administration. Intravenous vitamin C may also interfere with some glucose monitoring meters, causing either false-positive or false-negative results [135], so care should be taken with patients in vitamin C trials who have to monitor glucose levels [136–140]. Clinicians should also avoid using point-of-care glucose monitors to detect plasma vitamin C concentration in patients receiving ascorbate infusions [141] to avoid receiving inaccurate results [136–140].

PHARMACOLOGIC ASCORBATE SYNERGY WITH STANDARD CHEMOTHERAPY AND RADIOTHERAPY AGENTS

Many clinical trials provide evidence that pharmacologic ascorbate is well tolerated in patients with minimal toxicity, improving quality of life and reducing adverse events in combination with radiation or chemotherapy [66,101,102,129]. Some studies have described synergy between vitamin C and chemotherapeutics to reduce chemotherapy side effects [142–144]. A phase 1/2 study of patients with various advanced cancer

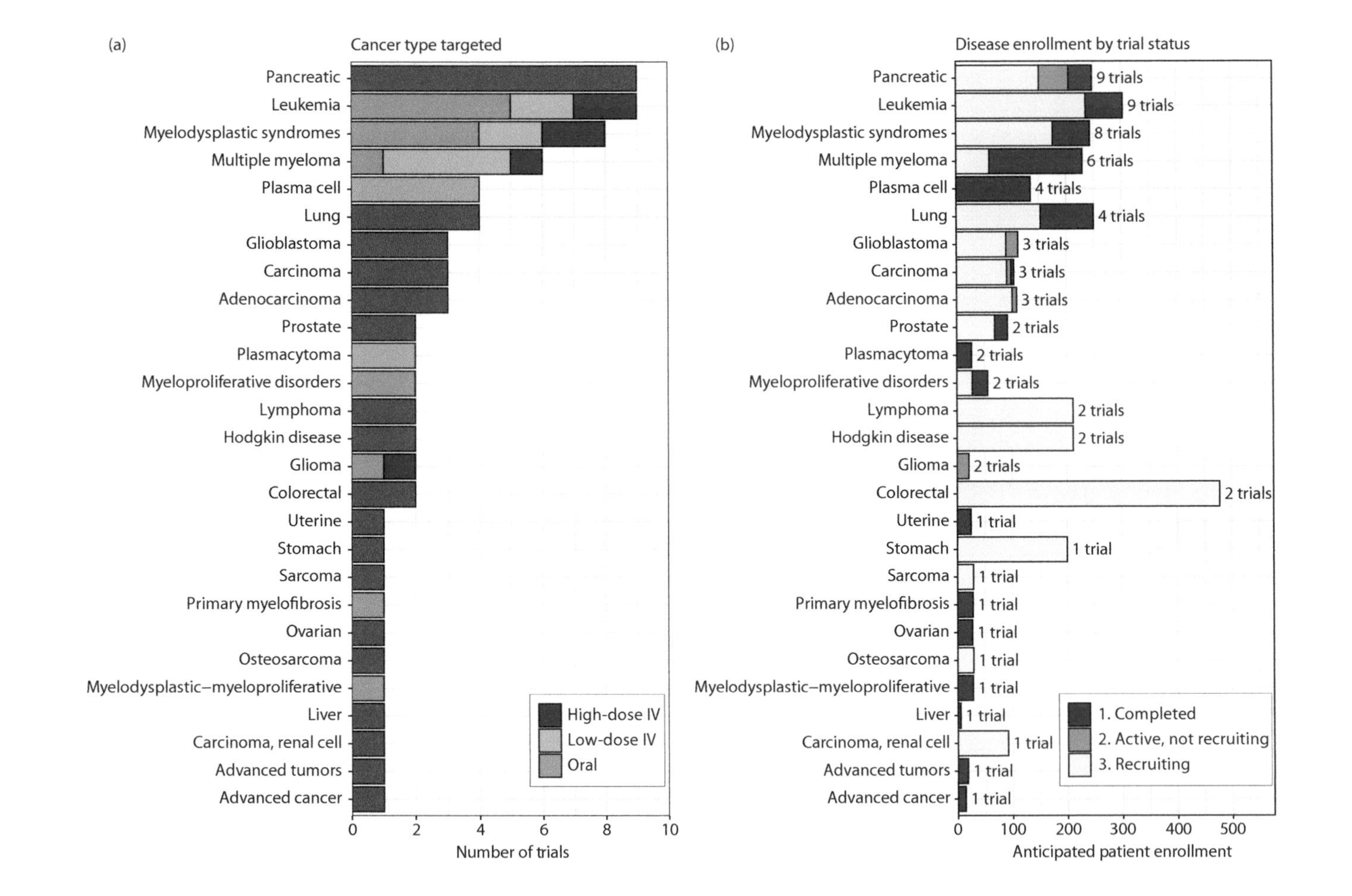

Figure 7.4. Cancer types investigated in vitamin C clinical trials by (a) intravenous [IV] or oral administration of vitamin C and by (b) trial status as of November 29, 2018. The clinicaltrials.gov label "condition_browse" was used to extract all cancer types studied in vitamin C clinical trials from ClinicalTrials.gov. Diseases were then manually grouped into 29 disease categories from an initial 53 diseases by the authors. (a) "Low-dose IV" was categorized as IV ascorbate dosing at 1 g daily and below. "High-dose IV" included any IV ascorbate dose above 1 g daily. (b) Cumulative actual and planned patient enrollment for vitamin C trials investigating each cancer type is given. "Completed" trials describe studies that have ended with participants no longer being treated. "Active, not recruiting" trials are ongoing, but new participants are not being enrolled. "Recruiting" trials are actively recruiting participants. For completed trials, actual (not planned) patient enrollment is depicted.

types treated with pharmacologic ascorbate and standard care chemotherapy found that 3 out of 14 patients experienced stable disease with unusually favorable clinical trajectories deemed highly unlikely to result from chemotherapy alone (probability less than 33%) [145]. The authors conclude that these three cases could be exceptional responders, and future trials might focus on the cancer types of these patients—cervical, biliary tract, and head and neck cancers [145]. A phase 1/2a study treated locally advanced or metastatic pancreatic cancer patients with pharmacologic ascorbate (dose escalated from 25 to 100 g) and gemcitabine (1000 mg/m^2) singularly for weeks 1–3 and together in week 4 [100]. Twelve participants completed phase 1, which involved a pharmacokinetic analysis of pharmacologic ascorbate and gemcitabine separately. The same 12 participants in phase 2a received combination therapy of pharmacologic ascorbate and gemcitabine three times a week for 2 consecutive weeks followed by a rest week until the patient progressed or withdrew from the study. Median overall survival was 15.1 months for the 12 participants: 5 did not survive over a year, 6 survived over one year, and 1 survived more than two decades after diagnosis [100]. The only adverse events related to pharmacologic ascorbate were grade 1 nausea and thirst.

In a 2017 phase 2 study of advanced stage, non-small cell lung cancer, participants were treated with a combination of carboplatin (AUC = 6, four cycles), paclitaxel (200 mg/m^2, four cycles), and pharmacologic ascorbate (two 75 g infusions per week, up to four cycles) [101]. No severe adverse events (grade 3 or 4) were attributed to pharmacologic ascorbate treatment. Among participants who completed the trial, nine had confirmed stable disease, four had partial response to treatment, and one had disease progression [101]. These clinical trials demonstrate that pharmacologic ascorbate is a nontoxic therapy that often reduces toxicity of chemotherapies when used in combination with standard treatments.

Pharmacologic ascorbate treatment has been controversial for patients undergoing chemotherapy because its electron donor activity has been proposed by some to reduce the effective level of reactive oxygen species–mediated cell injury induced during chemotherapy [146]. As mentioned earlier, at least at pharmacologic doses, ascorbate as an electron donor functions as a prodrug for generations of oxidant compounds, via formation of hydrogen peroxide as an oxidant precursor/generator. To our knowledge, there is no evidence showing that pharmacologic ascorbate interferes with chemotherapy; instead, it is often additive to or synergistic with chemotherapy. A report from *in vitro* studies suggested interference with chemotherapy, but investigators used dehydroascorbic acid, and not ascorbic acid [147,148]. A recent review reports a preponderance of evidence in favor of vitamin C use with chemotherapy [149].

One of the most commonly used chemotherapeutic agents in combination with ascorbate is gemcitabine (Figure 7.2a). The six agents most commonly used in combination with vitamin C are chemotherapy agents, with many trials actively recruiting participants (Figure 7.2b). Certain chemotherapy treatments cause antioxidant depletion and decreases in plasma ascorbate concentrations [150,151]. Hematologic cancer patients often undergo treatment with multiple chemotherapy agents, which significantly lowers vitamin C status [53,54,152–154]. Concurrent vitamin C treatment in patients receiving chemotherapy could replenish circulating vitamin C levels and decrease lipid oxidation [151,152]. In addition, some chemotherapy agents act through oxidative mechanisms. Because ascorbate in pharmacologic concentrations acts as a pro-oxidant, there could be synergy with chemotherapeutic agents acting via oxidative mechanisms. Conversely, there has been concern that secondary to its antioxidant properties, ascorbic acid may interfere with the efficacy of these chemotherapy agents. As a theoretical compromise, pharmacologic ascorbate is typically given the day before or after chemotherapy treatment to eliminate the possibility of interactions [128,155]. Vitamin C is also recommended to be administered five half-lives prior to chemotherapy treatment to eliminate potential interactions [156], but this could mask potential synergies with other pro-oxidant therapies. Since vitamin C is rapidly cleared by the kidneys with a half-life of less than 2 hours in circulation [95], possible treatment interaction may help to explain the observed radioprotective effects when radiation treatment was administered only 2 hours after intraperitoneal vitamin C administration in an animal model [157], although the administered ascorbate dose was lower than that typically found to slow tumor growth. Regarding radiation treatment, however, other models discussed later indicate that pharmacologic ascorbate can have ideal dual actions, as a radiosensitizer for tumor treatment with concurrent radioprotective effects on surrounding normal tissue [129,158].

Pharmacologic ascorbate may improve the efficacy of chemotherapy treatments (i.e., reduce the toxicity of chemotherapy or increase its efficacy) as demonstrated in *in vitro* and preclinical studies [40,159]. Ascorbate and carboplatin induced greater cell death in ovarian cancer cell lines (OVCAR5, OVCAR8, SHIN3) compared to either agent individually [66]. In some animal studies, pharmacologic ascorbate alone decreased tumor growth and promoted survival just as effectively as chemotherapeutic agents themselves [95]. In one particular mouse model of pancreatic cancer, combination treatment of gemcitabine and pharmacologic ascorbate decreased tumor volume compared to gemcitabine alone [46]. The combination treatment of vitamin C with various chemotherapy agents, such as gemcitabine, paclitaxel, carboplatin, melphalan, carfilzomib, bortezomib, cisplatin, and temozolomide, decreased xenograft tumor growth and increased survival [46,66,160,161]. Other animal studies demonstrated that pharmacologic ascorbate has similar anticancer effects compared to conventional chemotherapy or reduces toxicity from chemotherapy [46,66,161,162]. In some models there is no additive effect of combining chemotherapy with pharmacologic ascorbate compared to ascorbate alone. For example, in a murine melanoma model, antitumor activity from pharmacologic ascorbate was described, but minimal further tumor reduction was detected with addition of the chemotherapy agents dacarbazine and valproic acid [162].

While an increased number of randomized controlled trials using pharmacologic ascorbate treatment have opened, completed pharmacologic ascorbate studies were often not randomized controlled trials, so the efficacy of pharmacologic ascorbate has not been adequately determined. To date, two randomized pharmacologic vitamin C cancer clinical trials have been completed with reported results [66,163], and an additional randomized pharmacologic ascorbate trial with reported results [164]. In one of the completed trials of stage 3 and 4 ovarian cancer patients ($n = 27$) receiving carboplatin and paclitaxel chemotherapy, results included a trend toward overall survival, an 8.75-month increase in progression-free survival, and reduced adverse events in the pharmacologic ascorbate arm (75 or 100 g twice weekly for 12 months) [66]. The ascorbate group reported fewer grade 1 and 2 adverse events compared to the chemotherapy-only group. Participants in the ascorbate arm reported lower levels of toxicities frequently associated with carboplatin and paclitaxel, such as low-grade gastrointestinal toxicities, hepatobiliary toxicity, and pulmonary and renal toxicities [66]. The ascorbate arm had a higher median time to disease progression/relapse at 25.5 months compared to the chemotherapy-only arm (16.75 months) [66].

The possibility of ascorbic acid potentiating radiotherapy is promising and needs further research. Most but not all animal models [157] show the radiosensitizing effects of pharmacologic ascorbate synergistically acting with radiotherapy to decrease tumor growth [165,166] and increase survival [165]. In xenograft models of non-small cell lung cancer and glioblastoma multiforme, the triple combination of chemotherapy (carboplatin for lung cancer and temozolomide for glioblastoma), radiation therapy, and pharmacologic ascorbate produced prolonged survival compared to the dual combination of treatment of chemotherapy and radiation therapy [101]. Clinical findings were consistent with these results [101]. One hypothesis to explain the radiosensitizing effect of pharmacologic ascorbate is through decreasing clonogenic survival and double-stranded DNA breaks while simultaneously inhibiting radiation-induced damage in normal tissue [129]. Pharmacologic ascorbate's radiosensitizing effect and protection of surrounding tissues could have immense clinical value. It may be possible to utilize higher radiation doses to reduce tumor burden, while simultaneously decreasing radiation-induced adverse events, which are substantial and sometimes long term [129,167]. Further studies are needed to determine optimal timing and dosing of combined pharmacologic ascorbate with radiotherapy to evaluate synergy and radioprotection.

NEED FOR FUTURE RESEARCH AND PHARMACOLOGIC ASCORBATE CLINICAL TRIALS

New data are needed to elucidate the mechanisms of action of pharmacologic ascorbate to help in designing cancer clinical trials and to identify predictive biomarkers for patient selection and development of potent combination treatments. Based on preclinical studies, pharmacologic ascorbate may be more efficacious in tumors with the following mutations: *KRAS*, *BRAF*, *TET2*, *IDH1*, *IDH2*, *VHL*, *SH*, and *FDH* [168]. In addition, it is of key importance to determine appropriate dosing, timing, frequency, and duration of treatment. In a survey of alternative medicine practitioners, wide ranges of intravenous vitamin C dosing parameters were used: 1–200 g/infusion, 1–7 times per week, and 1–80 treatments [169]. It is difficult to interpret and compare data when extreme variability in dosing parameters is employed not only in clinical trial settings but throughout wider patient care environments. Combination treatment with vitamin C may also enhance sensitivity to particular anticancer agents, perhaps leading to reductions in drug dosage [78,170].

A key research priority is to determine appropriate dosing regimens for ascorbate, with or without combination drug treatment. There are only 29 trials in which high-dose ascorbate was utilized [19,66,99–102,104,128,129,145,163,164,171,172]. Compared to standard phase 3 trials, numbers of ascorbate-treated patients are small, but high-dose ascorbate treatment shows promising clinical outcomes. In these ascorbate trials, intravenous dosing was approximately 1 g/kg, with treatment two to three times weekly, continuing for a minimum of 8 weeks but much longer in some trials. By contrast, recently opened trials in myelodysplastic syndrome utilize much different dosing schemes, either continuous intravenous ascorbate for 5 days per month or with oral ascorbate alone. These dosing regimens stand in stark contrast to those in which outcomes are already available and promising. If the goal in myelodysplastic syndrome trials is to treat documented ascorbate deficiency, thereby enhancing TET actions, then the myelodysplastic syndrome trial designs have plausibility [77,78]. However, if the goals for myelodysplastic syndrome trials are based on pharmacologic use of ascorbate in enhancing TET actions, then their current designs lack supportive and foundational dosing data [78], and preclinical animal experiments would be informative. Unfortunately, potential danger lurks for the entire ascorbate treatment field, not just for myelodysplastic syndrome. If the myelodysplastic syndrome trials fail, many other trials with ascorbate may be at risk, especially those that utilize intravenous pharmacologic dosing schemes (i.e., 1 g/kg, two to three times weekly) where there is already some clinical basis for efficacy. We cannot emphasize enough that doses, duration, and frequency all do matter, and matter greatly [49,124]. We do not have the luxury of error, given the absence of pharmaceutical industry interest in pharmacologic ascorbate, the expense of clinical trials, and the past history/bias of ascorbate in cancer treatment.

Among the 42 trials analyzed, 13 were randomized and 4 included placebo controls. While safety and tolerability of pharmacologic ascorbate have been demonstrated in clinical trials, evidence of clinical efficacy is limited due to the lack of rigorous clinical trial data on the effect of pharmacologic ascorbate in certain cancer patient populations. In single-arm clinical trials of combined intravenous ascorbate treatment with standard therapies, it is unknown which treatment provided which effects [92]. Appropriately designed randomized placebo-controlled trials are needed to associate pharmacologic ascorbate with clinical benefit and/or toxicity and to help establish appropriate clinical guidelines. Primary endpoints could include progression-free survival, overall survival, and/or tumor response rate [49]. Secondary aims could investigate adverse events and quality of life, as well as predictive biomarkers. Vitamin C plasma concentration should also be measured at baseline in order to accurately measure the effect of vitamin C. Rigorous cancer clinical trial data are needed not only to further our understanding of ascorbate's efficacy in various settings, but also to potentially receive approval from the U.S. Food and Drug Administration for pharmacologic ascorbate treatment, which will better regulate its use as a cancer therapeutic in combination with other agents. Fortunately, randomized controlled trials to test antitumor effects of vitamin C are being developed and performed. The future of pharmacologic ascorbate therapy in the field of oncology is promising as long as phase 2 and 3 studies to investigate the efficacy of vitamin C in combination with standard cancer treatments are well designed and executed. Further, we must remain mindful of what we already know and what gaps exist regarding pharmacologic ascorbate dosing, frequency, and duration.

REFERENCES

1. Institute of Medicine. 2000. *Dietary Reference Intakes for Vitamin C, Vitamin E, Selenium, and Carotenoids*. The National Academies Press, Washington, DC.
2. Padayatty, S. J. and Levine, M. 2016. Vitamin C: The known and the unknown and Goldilocks. *Oral. Dis.* 22, 463–493.
3. Buettner, G. R. S., F.Q. 2004. Ascorbate as an antioxidant. In *Vitamin C: Function and Biochemistry in Animals and Plants* (Asard, H. M., James, M., Smirnoff, Nicholas, eds.). pp. 173–188, Garland Science/BIOS Scientific Publishers.
4. Levine, M., Padayatty, S. J. and Espey, M. G. 2011. Vitamin C: A concentration-function approach yields pharmacology and therapeutic discoveries. *Adv. Nutr.* 2, 78–88.
5. Mc, C. W. 1954. Cancer: the preconditioning factor in pathogenesis; a new etiologic approach. *Arch. Pediatr.* 71, 313–322.
6. Cameron, E. and Pauling, L. 1973. Ascorbic acid and the glycosaminoglycans. An orthomolecular approach to cancer and other diseases. *Oncology* 27, 181–192.
7. Cameron, E. and Pauling, L. 1974. The orthomolecular treatment of cancer. I. The role of ascorbic acid in host resistance. *Chem. Biol. Interact.* 9, 273–283.
8. Cameron, E. and Campbell, A. 1974. The orthomolecular treatment of cancer. II. Clinical trial of high-dose ascorbic acid supplements in advanced human cancer. *Chem. Biol. Interact.* 9, 285–315.
9. Cameron, E., Campbell, A. and Jack, T. 1975. The orthomolecular treatment of cancer. III. Reticulum cell sarcoma: Double complete regression induced by high-dose ascorbic acid therapy. *Chem. Biol. Interact.* 11, 387–393.
10. Cameron, E. and Pauling, L. 1976. Supplemental ascorbate in the supportive treatment of cancer: prolongation of survival times in terminal human cancer. *Proc. Natl. Acad. Sci. USA* 73, 3685–3689.
11. Cameron, E. and Pauling, L. 1978. Supplemental ascorbate in the supportive treatment of cancer: reevaluation of prolongation of survival times in terminal human cancer. *Proc. Natl. Acad. Sci. USA* 75, 4538–4542.
12. Moertel, C. G., Fleming, T. R., Creagan, E. T., Rubin, J., O'Connell, M. J. and Ames, M. M. 1985. High-dose vitamin C versus placebo in the treatment of patients with advanced cancer who have had no prior chemotherapy. A randomized double-blind comparison. *N. Engl. J. Med.* 312, 137–141.
13. Creagan, E. T., Moertel, C. G., O'Fallon, J. R., Schutt, A. J., O'Connell, M. J., Rubin, J. and Frytak, S. 1979. Failure of high-dose vitamin C (ascorbic acid) therapy to benefit patients with advanced cancer. A controlled trial. *N. Engl. J. Med.* 301, 687–690.
14. Jacobs, C., Hutton, B., Ng, T., Shorr, R. and Clemons, M. 2015. Is there a role for oral or intravenous ascorbate (vitamin C) in treating patients with cancer? A systematic review. *Oncologist* 20, 210–223.
15. Wittes, R. E. 1985. Vitamin C and cancer. *N. Engl. J. Med.* 312, 178–179.
16. Padayatty, S. J. and Levine, M. 2001. New insights into the physiology and pharmacology of vitamin C. *CMAJ*. 164, 353–355.
17. Padayatty, S. J., Sun, H., Wang, Y., Riordan, H. D., Hewitt, S. M., Katz, A., Wesley, R. A. and Levine, M. (2004). Vitamin C pharmacokinetics: Implications for oral and intravenous use. *Ann. Intern. Med.* 140, 533–537.
18. Padayatty, S. J. and Levine, M. 2000. Reevaluation of ascorbate in cancer treatment: Emerging evidence, open minds and serendipity. *J. Am. Coll. Nutr.* 19, 423–425.
19. Hoffer, L. J., Levine, M., Assouline, S., Melnychuk, D., Padayatty, S. J., Rosadiuk, K., Rousseau, C., Robitaille, L. and Miller, W. H., Jr. 2008. Phase I clinical trial of i.v. ascorbic acid in advanced malignancy. *Ann. Oncol.* 19, 1969–1974.
20. Park, C. H., Kimler, B. F., Yi, S. Y., Park, S. H., Kim, K., Jung, C. W., Kim, S. H. et al. 2009. Depletion of L-ascorbic acid alternating with its supplementation in the treatment of patients with acute myeloid leukemia or myelodysplastic syndromes. *Eur. J. Haematol.* 83, 108–118.
21. Young, J. I., Zuchner, S. and Wang, G. 2015. Regulation of the epigenome by vitamin C. *Annu. Rev. Nutr.* 35, 545–564.
22. Levine, M., Rumsey, S. C., Daruwala, R., Park, J. B. and Wang, Y. 1999. Criteria and recommendations for vitamin C intake. *JAMA* 281, 1415–1423.
23. Padayatty, S. J. and Levine, M. 2009. Antioxidant supplements and cardiovascular disease in men. *JAMA*. 301, 1336; author reply 1336–1337.
24. Levine, M., Conry-Cantilena, C., Wang, Y., Welch, R. W., Washko, P. W., Dhariwal, K. R., Park, J. B. et al. 1996. Vitamin C

pharmacokinetics in healthy volunteers: Evidence for a recommended dietary allowance. *Proc. Natl. Acad. Sci. USA* 93, 3704–3709.
25. Levine, M., Wang, Y., Padayatty, S. J. and Morrow, J. 2001. A new recommended dietary allowance of vitamin C for healthy young women. *Proc. Natl. Acad. Sci. USA* 98, 9842–9846.
26. Leung, P. Y., Miyashita, K., Young, M. and Tsao, C. S. 1993. Cytotoxic effect of ascorbate and its derivatives on cultured malignant and nonmalignant cell lines. *Anticancer Res.* 13, 475–480.
27. Bram, S., Froussard, P., Guichard, M., Jasmin, C., Augery, Y., Sinoussi-Barre, F. and Wray, W. 1980. Vitamin C preferential toxicity for malignant melanoma cells. *Nature* 284, 629–631.
28. Casciari, J. J., Riordan, N. H., Schmidt, T. L., Meng, X. L., Jackson, J. A. and Riordan, H. D. 2001. Cytotoxicity of ascorbate, lipoic acid, and other antioxidants in hollow fibre in vitro tumours. *Br. J. Cancer.* 84, 1544–1550.
29. Chen, Q., Espey, M. G., Krishna, M. C., Mitchell, J. B., Corpe, C. P., Buettner, G. R., Shacter, E. and Levine, M. 2005. Pharmacologic ascorbic acid concentrations selectively kill cancer cells: action as a pro-drug to deliver hydrogen peroxide to tissues. *Proc. Natl. Acad. Sci. USA* 102, 13604–13609.
30. Fujii, H., Amano, O., Kochi, M. and Sakagami, H. 2003. Mitochondrial control of cell death induction by sodium 5,6-benzylidene-L-ascorbate. *Anticancer Res.* 23, 1353–1356.
31. Jamison, J. M., Gilloteaux, J., Nassiri, M. R., Venugopal, M., Neal, D. R. and Summers, J. L. 2004. Cell cycle arrest and autoschizis in a human bladder carcinoma cell line following vitamin C and vitamin K3 treatment. *Biochem. Pharmacol.* 67, 337–351.
32. Makino, Y., Sakagami, H. and Takeda, M. 1999. Induction of cell death by ascorbic acid derivatives in human renal carcinoma and glioblastoma cell lines. *Anticancer Res.* 19, 3125–3132.
33. Maramag, C., Menon, M., Balaji, K. C., Reddy, P. G. and Laxmanan, S. 1997. Effect of vitamin C on prostate cancer cells in vitro: Effect on cell number, viability, and DNA synthesis. *Prostate* 32, 188–195.
34. Munkres, K. D. 1979. Ageing of Neurospora crassa. VIII. Lethality and mutagenicity of ferrous ions, ascorbic acid, and malondialdehyde. *Mech. Ageing Dev.* 10, 249–260.
35. Chen, Q., Espey, M. G., Sun, A. Y., Pooput, C., Kirk, K. L., Krishna, M. C., Khosh, D. B., Drisko, J. and Levine, M. 2008. Pharmacologic doses of ascorbate act as a prooxidant and decrease growth of aggressive tumor xenografts in mice. *Proc. Natl. Acad. Sci. USA* 105, 11105–11109.
36. Chen, Q., Espey, M. G., Sun, A. Y., Lee, J. H., Krishna, M. C., Shacter, E., Choyke, P. L. 2007. Ascorbate in pharmacologic concentrations selectively generates ascorbate radical and hydrogen peroxide in extracellular fluid in vivo. *Proc. Natl. Acad. Sci. USA* 104, 8749–8754.
37. Chen, P., Yu, J., Chalmers, B., Drisko, J., Yang, J., Li, B. and Chen, Q. 2012. Pharmacological ascorbate induces cytotoxicity in prostate cancer cells through ATP depletion and induction of autophagy. *Anticancer Drugs* 23, 437–444.
38. Chen, P., Stone, J., Sullivan, G., Drisko, J. A. and Chen, Q. 2011. Anti-cancer effect of pharmacologic ascorbate and its interaction with supplementary parenteral glutathione in preclinical cancer models. *Free Radic. Biol. Med.* 51, 681–687.
39. Verrax, J. and Calderon, P. B. 2009. Pharmacologic concentrations of ascorbate are achieved by parenteral administration and exhibit antitumoral effects. *Free Radic. Biol. Med.* 47, 32–40.
40. Fromberg, A., Gutsch, D., Schulze, D., Vollbracht, C., Weiss, G., Czubayko, F. and Aigner, A. 2011. Ascorbate exerts anti-proliferative effects through cell cycle inhibition and sensitizes tumor cells towards cytostatic drugs. *Cancer Chemother. Pharmacol.* 67, 1157–1166.
41. Pathi, S. S., Lei, P., Sreevalsan, S., Chadalapaka, G., Jutooru, I. and Safe, S. 2011. Pharmacologic doses of ascorbic acid repress specificity protein (Sp) transcription factors and Sp-regulated genes in colon cancer cells. *Nutr. Cancer* 63, 1133–1142.
42. Lin, Z. Y. and Chuang, W. L. 2010. Pharmacologic concentrations of ascorbic acid cause diverse influence on differential expressions of angiogenic chemokine genes in different hepatocellular carcinoma cell lines. *Biomed. Pharmacother.* 64, 348–351.
43. Takemura, Y., Satoh, M., Satoh, K., Hamada, H., Sekido, Y. and Kubota, S. 2010. High dose of ascorbic acid induces cell death in mesothelioma cells. *Biochem. Biophys. Res. Commun.* 394, 249–253.
44. Hardaway, C. M., Badisa, R. B. and Soliman, K. F. 2012. Effect of ascorbic acid and hydrogen peroxide on mouse neuroblastoma cells. *Mol. Med. Rep.* 5, 1449–1452.

45. Du, J., Martin, S. M., Levine, M., Wagner, B. A., Buettner, G. R., Wang, S. H., Taghiyev, A. F., Du, C., Knudson, C. M. and Cullen, J. J. 2010. Mechanisms of ascorbate-induced cytotoxicity in pancreatic cancer. *Clin. Cancer Res.* 16, 509–520.
46. Espey, M. G., Chen, P., Chalmers, B., Drisko, J., Sun, A. Y., Levine, M. and Chen, Q. 2011. Pharmacologic ascorbate synergizes with gemcitabine in preclinical models of pancreatic cancer. *Free Radic. Biol. Med.* 50, 1610–1619.
47. Pollard, H. B., Levine, M. A., Eidelman, O. and Pollard, M. 2010. Pharmacological ascorbic acid suppresses syngeneic tumor growth and metastases in hormone-refractory prostate cancer. *In Vivo.* 24, 249–255.
48. Yeom, C. H., Lee, G., Park, J. H., Yu, J., Park, S., Yi, S. Y., Lee, H. R., Hong, Y. S., Yang, J. and Lee, S. 2009. High dose concentration administration of ascorbic acid inhibits tumor growth in BALB/C mice implanted with sarcoma 180 cancer cells via the restriction of angiogenesis. *J. Transl. Med.* 7, 70.
49. Shenoy, N., Creagan, E., Witzig, T. and Levine, M. 2018. Ascorbic acid in cancer treatment: Let the phoenix fly. *Cancer Cell* 34, 700–706.
50. Kawada, H., Kaneko, M., Sawanobori, M., Uno, T., Matsuzawa, H., Nakamura, Y., Matsushita, H. and Ando, K. 2013. High concentrations of L-ascorbic acid specifically inhibit the growth of human leukemic cells via downregulation of HIF-1alpha transcription. *PLoS One.* 8, e62717.
51. Kuiper, C., Dachs, G. U., Currie, M. J. and Vissers, M. C. 2014. Intracellular ascorbate enhances hypoxia-inducible factor (HIF)-hydroxylase activity and preferentially suppresses the HIF-1 transcriptional response. *Free Radic. Biol. Med.* 69, 308–317.
52. Kuiper, C., Molenaar, I. G., Dachs, G. U., Currie, M. J., Sykes, P. H. and Vissers, M. C. 2010. Low ascorbate levels are associated with increased hypoxia-inducible factor-1 activity and an aggressive tumor phenotype in endometrial cancer. *Cancer Res.* 70, 5749–5758.
53. Huijskens, M. J., Wodzig, W. K., Walczak, M., Germeraad, W. T. and Bos, G. M. 2016. Ascorbic acid serum levels are reduced in patients with hematological malignancies. *Results Immunol.* 6, 8–10.
54. Liu, M., Ohtani, H., Zhou, W., Orskov, A. D., Charlet, J., Zhang, Y. W., Shen, H. et al. 2016. Vitamin C increases viral mimicry induced by 5-aza-2′-deoxycytidine. *Proc. Natl. Acad. Sci. USA* 113, 10238–10244.
55. Mayland, C. R., Bennett, M. I. and Allan, K. 2005. Vitamin C deficiency in cancer patients. *Palliat. Med.* 19, 17–20.
56. Kuiper, C., Vissers, M. C. and Hicks, K. O. 2014. Pharmacokinetic modeling of ascorbate diffusion through normal and tumor tissue. *Free Radic. Biol. Med.* 77, 340–352.
57. Rawal, M., Schroeder, S. R., Wagner, B. A., Cushing, C. M., Welsh, J. L., Button, A. M., Du, J., Sibenaller, Z. A., Buettner, G. R. and Cullen, J. J. 2013. Manganoporphyrins increase ascorbate-induced cytotoxicity by enhancing H2O2 generation. *Cancer Res.* 73, 5232–5241.
58. Buettner, G. R. and Jurkiewicz, B. A. 1996. Catalytic metals, ascorbate and free radicals: Combinations to avoid. *Radiat. Res.* 145, 532–541.
59. Vissers, M. C. M. and Das, A. B. 2018. Potential mechanisms of action for vitamin C in cancer: Reviewing the evidence. *Front. Physiol.* 9, 809.
60. Doskey, C. M., Buranasudja, V., Wagner, B. A., Wilkes, J. G., Du, J., Cullen, J. J. and Buettner, G. R. 2016. Tumor cells have decreased ability to metabolize H_2O_2: Implications for pharmacological ascorbate in cancer therapy. *Redox Biol.* 10, 274–284.
61. Du, J., Wagner, B. A., Buettner, G. R. and Cullen, J. J. 2015. Role of labile iron in the toxicity of pharmacological ascorbate. *Free Radic. Biol. Med.* 84, 289–295.
62. Sestili, P., Brandi, G., Brambilla, L., Cattabeni, F. and Cantoni, O. 1996. Hydrogen peroxide mediates the killing of U937 tumor cells elicited by pharmacologically attainable concentrations of ascorbic acid: Cell death prevention by extracellular catalase or catalase from cocultured erythrocytes or fibroblasts. *J. Pharmacol. Exp. Ther.* 277, 1719–1725.
63. Ranzato, E., Biffo, S. and Burlando, B. 2011. Selective ascorbate toxicity in malignant mesothelioma: a redox Trojan mechanism. *Am. J. Respir. Cell Mol. Biol.* 44, 108–117.
64. Klingelhoeffer, C., Kammerer, U., Koospal, M., Muhling, B., Schneider, M., Kapp, M., Kubler, A., Germer, C. T. and Otto, C. 2012. Natural resistance to ascorbic acid induced oxidative stress is mainly mediated by catalase activity in human cancer cells and catalase-silencing sensitizes to oxidative stress. *BMC Complement. Altern. Med.* 12, 61.
65. Shatzer, A. N., Espey, M. G., Chavez, M., Tu, H., Levine, M. and Cohen, J. I. 2013. Ascorbic acid kills Epstein-Barr virus positive Burkitt

lymphoma cells and Epstein-Barr virus transformed B-cells in vitro, but not in vivo. *Leuk. Lymphoma* 54, 1069–1078.

66. Ma, Y., Chapman, J., Levine, M., Polireddy, K., Drisko, J. and Chen, Q. 2014. High-dose parenteral ascorbate enhanced chemosensitivity of ovarian cancer and reduced toxicity of chemotherapy. *Sci. Transl. Med.* 6, 222ra218.
67. Huang, Y. C., Chang, T. K., Fu, Y. C. and Jan, S. L. 2014. C for colored urine: acute hemolysis induced by high-dose ascorbic acid. *Clin. Toxicol. (Phila).* 52, 984.
68. Rees, D. C., Kelsey, H. and Richards, J. D. 1993. Acute haemolysis induced by high dose ascorbic acid in glucose-6-phosphate dehydrogenase deficiency. *BMJ.* 306, 841–842.
69. Levine, M. and Violet, P. C. 2017. Data triumph at C. *Cancer Cell* 31, 467–469.
70. Ohno, S., Ohno, Y., Suzuki, N., Soma, G. and Inoue, M. 2009. High-dose vitamin C (ascorbic acid) therapy in the treatment of patients with advanced cancer. *Anticancer Res.* 29, 809–815.
71. Yun, J., Mullarky, E., Lu, C., Bosch, K. N., Kavalier, A., Rivera, K., Roper, J. et al. 2015. Vitamin C selectively kills KRAS and BRAF mutant colorectal cancer cells by targeting GAPDH. *Science* 350, 1391–1396.
72. van der Reest, J. and Gottlieb, E. 2016. Anti-cancer effects of vitamin C revisited. *Cell Res.* 26, 269–270.
73. Ma, E., Chen, P., Wilkins, H. M., Wang, T., Swerdlow, R. H. and Chen, Q. 2017. Pharmacologic ascorbate induces neuroblastoma cell death by hydrogen peroxide mediated DNA damage and reduction in cancer cell glycolysis. *Free Radic. Biol. Med.* 113, 36–47.
74. Ye, M., Pang, N., Wan, T., Huang, Y., Wei, T., Jiang, X., Zhou, Y. et al. 2019. Oxidized vitamin C (DHA) overcomes resistance to EGFR-targeted therapy of lung cancer through disturbing energy homeostasis. *J. Cancer* 10, 757–764.
75. Blaschke, K., Ebata, K. T., Karimi, M. M., Zepeda-Martinez, J. A., Goyal, P., Mahapatra, S., Tam, A. et al. 2013. Vitamin C induces TET-dependent DNA demethylation and a blastocyst-like state in ES cells. *Nature* 500, 222–226.
76. Yin, R., Mao, S. Q., Zhao, B., Chong, Z., Yang, Y., Zhao, C., Zhang, D. et al. 2013. Ascorbic acid enhances TET-mediated 5-methylcytosine oxidation and promotes DNA demethylation in mammals. *J. Am. Chem. Soc.* 135, 10396–10403.
77. Agathocleous, M., Meacham, C. E., Burgess, R. J., Piskounova, E., Zhao, Z., Crane, G. M., Cowin, B. L. et al. 2017. Ascorbate regulates haematopoietic stem cell function and leukaemogenesis. *Nature* 549, 476–481.
78. Cimmino, L., Dolgalev, I., Wang, Y., Yoshimi, A., Martin, G. H., Wang, J., Ng, V. et al. 2017. Restoration of TET2 function blocks aberrant self-renewal and leukemia progression. *Cell.* 170, 1079–1095, e1020.
79. Shenoy, N., Bhagat, T., Nieves, E., Stenson, M., Lawson, J., Choudhary, G. S., Habermann, T. et al. 2017. Upregulation of TET activity with ascorbic acid induces epigenetic modulation of lymphoma cells. *Blood Cancer J.* 7, e587.
80. Gustafson, C. B., Yang, C., Dickson, K. M., Shao, H., Van Booven, D., Harbour, J. W., Liu, Z. J. and Wang, G. 2015. Epigenetic reprogramming of melanoma cells by vitamin C treatment. *Clin. Epigenetics.* 7, 51.
81. Wu, X. and Zhang, Y. 2017. TET-mediated active DNA demethylation: mechanism, function and beyond. *Nat. Rev. Genet.* 18, 517–534.
82. Guillamot, M., Cimmino, L. and Aifantis, I. 2016. The impact of DNA methylation in hematopoietic malignancies. *Trends Cancer* 2, 70–83.
83. Bejar, R., Lord, A., Stevenson, K., Bar-Natan, M., Perez-Ladaga, A., Zaneveld, J., Wang, H. et al. 2014. TET2 mutations predict response to hypomethylating agents in myelodysplastic syndrome patients. *Blood* 124, 2705–2712.
84. Figueroa, M. E., Abdel-Wahab, O., Lu, C., Ward, P. S., Patel, J., Shih, A., Li, Y. et al. 2010. Leukemic IDH1 and IDH2 mutations result in a hypermethylation phenotype, disrupt TET2 function, and impair hematopoietic differentiation. *Cancer Cell* 18, 553–567.
85. Hu, C. Y., Mohtat, D., Yu, Y., Ko, Y. A., Shenoy, N., Bhattacharya, S., Izquierdo, M. C. et al. 2014. Kidney cancer is characterized by aberrant methylation of tissue-specific enhancers that are prognostic for overall survival. *Clin. Cancer Res.* 20, 4349–4360.
86. Jiang, Y., Dunbar, A., Gondek, L. P., Mohan, S., Rataul, M., O'Keefe, C., Sekeres, M., Saunthararajah, Y. and Maciejewski, J. P. 2009. Aberrant DNA methylation is a dominant mechanism in MDS progression to AML. *Blood* 113, 1315–1325.

87. Letouze, E., Martinelli, C., Loriot, C., Burnichon, N., Abermil, N., Ottolenghi, C., Janin, M. et al. 2013. SDH mutations establish a hypermethylator phenotype in paraganglioma. *Cancer Cell* 23, 739–752.
88. Patnaik, M. M. and Tefferi, A. 2016. Chronic myelomonocytic leukemia: 2016 update on diagnosis, risk stratification, and management. *Am. J. Hematol.* 91, 631–642.
89. Rasmussen, K. D., Jia, G., Johansen, J. V., Pedersen, M. T., Rapin, N., Bagger, F. O., Porse, B. T., Bernard, O. A., Christensen, J. and Helin, K. 2015. Loss of TET2 in hematopoietic cells leads to DNA hypermethylation of active enhancers and induction of leukemogenesis. *Genes Dev.* 29, 910–922.
90. Shenoy, N., Bhagat, T. D., Cheville, J., Lohse, C., Bhattacharyya, S., Tischer, A., Machha, V. et al. 2019. Ascorbic acid-induced TET activation mitigates adverse hydroxymethylcytosine loss in renal cell carcinoma. *J. Clin. Invest.* 130, 1612–1625.
91. Results can be found at: sysrev.com/p/6737. Normalized drug names were extracted from drugbank.ca. This normalization process was conducted through human review (sysrev.com/p/6963) of drug names and manual matching to controlled name given on drugbank.ca.
92. Nauman, G., Gray, J. C., Parkinson, R., Levine, M. and Paller, C. J. 2018. Systematic review of intravenous ascorbate in cancer clinical trials. *Antioxidants (Basel)*. 7.
93. Murray, M. T. 1996. *Encyclopedia of Nutritional Supplements: The Essential Guide for Improving Your Health Naturally*. Prima Publishing, Rocklin, CA.
94. McGuff Pharmaceuticals. 2017. ASCOR (Ascorbin Acid Injection, USP) full prescribing information: Contents. ed.)⊥eds.).
95. Carr, A. C. and Cook, J. 2018. Intravenous vitamin C for cancer therapy—Identifying the current gaps in our knowledge. *Front. Physiol.* 9, 1182.
96. Langemann, H., Torhorst, J., Kabiersch, A., Krenger, W. and Honegger, C. G. 1989. Quantitative determination of water- and lipid-soluble antioxidants in neoplastic and non-neoplastic human breast tissue. *Int. J. Cancer.* 43, 1169–1173.
97. Honegger, C. G., Torhorst, J., Langemann, H., Kabiersch, A. and Krenger, W. 1988. Quantitative determination of water-soluble scavengers in neoplastic and non-neoplastic human breast tissue. *Int. J. Cancer.* 41, 690–694.
98. Agus, D. B., Vera, J. C. and Golde, D. W. 1999. Stromal cell oxidation: A mechanism by which tumors obtain vitamin C. *Cancer Res.* 59, 4555–4558.
99. Monti, D. A., Mitchell, E., Bazzan, A. J., Littman, S., Zabrecky, G., Yeo, C. J., Pillai, M. V., Newberg, A. B., Deshmukh, S. and Levine, M. 2012. Phase I evaluation of intravenous ascorbic acid in combination with gemcitabine and erlotinib in patients with metastatic pancreatic cancer. *PLoS One* 7, e29794.
100. Polireddy, K., Dong, R., Reed, G., Yu, J., Chen, P., Williamson, S., Violet, P. C. et al. 2017. High dose parenteral ascorbate inhibited pancreatic cancer growth and metastasis: Mechanisms and a phase I/IIa study. *Sci. Rep.* 7, 17188.
101. Schoenfeld, J. D., Sibenaller, Z. A., Mapuskar, K. A., Wagner, B. A., Cramer-Morales, K. L., Furqan, M., Sandhu, S. et al. 2017. O_2(-) and H_2O_2-mediated disruption of Fe metabolism causes the differential susceptibility of NSCLC and GBM cancer cells to pharmacological ascorbate. *Cancer Cell* 31, 487–500, e488.
102. Welsh, J. L., Wagner, B. A., van't Erve, T. J., Zehr, P. S., Berg, D. J., Halfdanarson, T. R., Yee, N. S. et al. 2013. Pharmacological ascorbate with gemcitabine for the control of metastatic and node-positive pancreatic cancer (PACMAN): results from a phase I clinical trial. *Cancer Chemother. Pharmacol.* 71, 765–775.
103. Drisko, J. A., Serrano, O. K., Spruce, L. R., Chen, Q. and Levine, M. 2018. Treatment of pancreatic cancer with intravenous vitamin C: A case report. *Anticancer Drugs* 29, 373–379.
104. Nielsen, T. K., Hojgaard, M., Andersen, J. T., Jorgensen, N. R., Zerahn, B., Kristensen, B., Henriksen, T., Lykkesfeldt, J., Mikines, K. J. and Poulsen, H. E. 2017. Weekly ascorbic acid infusion in castration-resistant prostate cancer patients: A single-arm phase II trial. *Transl. Androl. Urol.* 6, 517–528.
105. Rouleau, L., Antony, A. N., Bisetto, S., Newberg, A., Doria, C., Levine, M., Monti, D. A. and Hoek, J. B. 2016. Synergistic effects of ascorbate and sorafenib in hepatocellular carcinoma: New insights into ascorbate cytotoxicity. *Free Radic. Biol. Med.* 95, 308–322.
106. Campbell, E. J., Vissers, M. C. M., Wohlrab, C., Hicks, K. O., Strother, R. M., Bozonet, S. M., Robinson, B. A. and Dachs, G. U. 2016. Pharmacokinetic and anti-cancer properties of high dose ascorbate in solid tumours of ascorbate-dependent mice. *Free Radic. Biol. Med.* 99, 451–462.

107. Cameron, E. 1991. Protocol for the use of vitamin C in the treatment of cancer. *Med. Hypotheses*. 36, 190–194.
108. Gonzalez, M. J., Miranda Massari, J. R., Duconge, J., Riordan, N. H. and Ichim, T. 2012. Schedule dependence in cancer therapy: Intravenous vitamin C and the systemic saturation hypothesis. *J. Orthomol. Med*. 27, 9–12.
109. Qazilbash, M. H., Saliba, R. M., Nieto, Y., Parikh, G., Pelosini, M., Khan, F. B., Jones, R. B. et al. 2008. Arsenic trioxide with ascorbic acid and high-dose melphalan: results of a phase II randomized trial. *Biol. Blood Marrow. Transplant*. 14, 1401–1407.
110. Bahlis, N. J., McCafferty-Grad, J., Jordan-McMurry, I., Neil, J., Reis, I., Kharfan-Dabaja, M., Eckman, J. et al. 2002. Feasibility and correlates of arsenic trioxide combined with ascorbic acid-mediated depletion of intracellular glutathione for the treatment of relapsed/refractory multiple myeloma. *Clin. Cancer Res*. 8, 3658–3668.
111. Welch, J. S., Klco, J. M., Gao, F., Procknow, E., Uy, G. L., Stockerl-Goldstein, K. E., Abboud, C. N. et al. 2011. Combination decitabine, arsenic trioxide, and ascorbic acid for the treatment of myelodysplastic syndrome and acute myeloid leukemia: A phase I study. *Am. J. Hematol*. 86, 796–800.
112. Sharma, M., Khan, H., Thall, P. F., Orlowski, R. Z., Bassett, R. L., Jr, Shah, N., Bashir, Q. et al. 2012. A randomized phase 2 trial of a preparative regimen of bortezomib, high-dose melphalan, arsenic trioxide, and ascorbic acid. *Cancer* 118, 2507–2515.
113. Bejanyan, N., Tiu, R. V., Raza, A., Jankowska, A., Kalaycio, M., Advani, A., Chan, J. et al. 2012. A phase 2 trial of combination therapy with thalidomide, arsenic trioxide, dexamethasone, and ascorbic acid (TADA) in patients with overlap myelodysplastic/myeloproliferative neoplasms (MDS/MPN) or primary myelofibrosis (PMF). *Cancer* 118, 3968–3976.
114. Berenson, J. R., Boccia, R., Siegel, D., Bozdech, M., Bessudo, A., Stadtmauer, E., Talisman Pomeroy, J. et al. 2006. Efficacy and safety of melphalan, arsenic trioxide and ascorbic acid combination therapy in patients with relapsed or refractory multiple myeloma: A prospective, multicentre, phase II, single-arm study. *Br. J. Haematol*. 135, 174–183.
115. Berenson, J. R., Matous, J., Swift, R. A., Mapes, R., Morrison, B. and Yeh, H. S. 2007. A phase I/II study of arsenic trioxide/bortezomib/ascorbic acid combination therapy for the treatment of relapsed or refractory multiple myeloma. *Clin. Cancer Res*. 13, 1762–1768.
116. Abou-Jawde, R. M., Reed, J., Kelly, M., Walker, E., Andresen, S., Baz, R., Karam, M. A. and Hussein, M. 2006. Efficacy and safety results with the combination therapy of arsenic trioxide, dexamethasone, and ascorbic acid in multiple myeloma patients: A phase 2 trial. *Med. Oncol*. 23, 263–272.
117. Chang, J. E., Voorhees, P. M., Kolesar, J. M., Ahuja, H. G., Sanchez, F. A., Rodriguez, G. A., Kim, K., Werndli, J., Bailey, H. H. and Kahl, B. S. 2009. Phase II study of arsenic trioxide and ascorbic acid for relapsed or refractory lymphoid malignancies: A Wisconsin Oncology Network study. *Hematol. Oncol*. 27, 11–16.
118. Bael, T. E., Peterson, B. L. and Gollob, J. A. 2008. Phase II trial of arsenic trioxide and ascorbic acid with temozolomide in patients with metastatic melanoma with or without central nervous system metastases. *Melanoma. Res*. 18, 147–151.
119. Subbarayan, P. R., Lima, M. and Ardalan, B. 2007. Arsenic trioxide/ascorbic acid therapy in patients with refractory metastatic colorectal carcinoma: A clinical experience. *Acta Oncol*. 46, 557–561.
120. Wu, K. L., Beksac, M., van Droogenbroeck, J., Amadori, S., Zweegman, S. and Sonneveld, P. 2006. Phase II multicenter study of arsenic trioxide, ascorbic acid and dexamethasone in patients with relapsed or refractory multiple myeloma. *Haematologica* 91, 1722–1723.
121. Held, L. A., Rizzieri, D., Long, G. D., Gockerman, J. P., Diehl, L. F., de Castro, C. M., Moore, J. O., Horwitz, M. E., Chao, N. J. and Gasparetto, C. 2013. A phase I study of arsenic trioxide (Trisenox), ascorbic acid, and bortezomib (Velcade) combination therapy in patients with relapsed/refractory multiple myeloma. *Cancer Invest*. 31, 172–176.
122. Fritz, H., Flower, G., Weeks, L., Cooley, K., Callachan, M., McGowan, J., Skidmore, B., Kirchner, L. and Seely, D. 2014. Intravenous vitamin C and cancer: A systematic review. *Integr. Cancer Ther*. 13, 280–300.
123. Grad, J. M., Bahlis, N. J., Reis, I., Oshiro, M. M., Dalton, W. S. and Boise, L. H. 2001. Ascorbic acid enhances arsenic trioxide-induced cytotoxicity in multiple myeloma cells. *Blood* 98, 805–813.
124. Violet, P. C. and Levine, M. 2017. Pharmacologic ascorbate in myeloma treatment: Doses matter. *EBioMedicine* 18, 9–10.

125. Berenson, J. R., Yellin, O., Woytowitz, D., Flam, M. S., Cartmell, A., Patel, R., Duvivier, H. et al. 2009. Bortezomib, ascorbic acid and melphalan (BAM) therapy for patients with newly diagnosed multiple myeloma: An effective and well-tolerated frontline regimen. *Eur. J. Haematol.* 82, 433–439.
126. Klein, F. 2017. Vitamin C wichtiges Kotherapeutikum? *Info Onkologie* 20, 48.
127. Wang, F., He, M. M., Wang, Z. X., Li, S., Jin, Y., Ren, C., Shi, S. M. et al. 2019. Phase I study of high-dose ascorbic acid with mFOLFOX6 or FOLFIRI in patients with metastatic colorectal cancer or gastric cancer. *BMC Cancer* 19, 460.
128. Stephenson, C. M., Levin, R. D., Spector, T. and Lis, C. G. 2013. Phase I clinical trial to evaluate the safety, tolerability, and pharmacokinetics of high-dose intravenous ascorbic acid in patients with advanced cancer. *Cancer Chemother. Pharmacol.* 72, 139–146.
129. Alexander, M. S., Wilkes, J. G., Schroeder, S. R., Buettner, G. R., Wagner, B. A., Du, J., Gibson-Corely, K. et al. 2018. Pharmacological ascorbate reduces radiation-induced normal tissue toxicity and enhances tumor radiosensitization in pancreatic cancer. *Cancer Res.* 78, 6838–6851.
130. Robitaille, L., Mamer, O. A., Miller, W. H., Jr, Levine, M., Assouline, S., Melnychuk, D., Rousseau, C. and Hoffer, L. J. 2009. Oxalic acid excretion after intravenous ascorbic acid administration. *Metabolism* 58, 263–269.
131. Lawton, J. M., Conway, L. T., Crosson, J. T., Smith, C. L. and Abraham, P. A. 1985. Acute oxalate nephropathy after massive ascorbic acid administration. *Arch. Intern. Med.* 145, 950–951.
132. Wong, K., Thomson, C., Bailey, R. R., McDiarmid, S. and Gardner, J. 1994. Acute oxalate nephropathy after a massive intravenous dose of vitamin C. *Aust. N. Z. J. Med.* 24, 410–411.
133. Cossey, L. N., Rahim, F. and Larsen, C. P. 2013. Oxalate nephropathy and intravenous vitamin C. *Am. J. Kidney Dis.* 61, 1032–1035.
134. Quinn, J., Gerber, B., Fouche, R., Kenyon, K., Blom, Z. and Muthukanagaraj, P. 2017. Effect of high-dose vitamin C infusion in a glucose-6-phosphate dehydrogenase-deficient patient. *Case Rep. Med.* 2017, 5202606.
135. Tang, Z., Du, X., Louie, R. F. and Kost, G. J. 2000. Effects of drugs on glucose measurements with handheld glucose meters and a portable glucose analyzer. *Am. J. Clin. Pathol.* 113, 75–86.
136. Cho, J., Ahn, S., Yim, J., Cheon, Y., Jeong, S. H., Lee, S. G. and Kim, J. H. 2016. Influence of vitamin C and maltose on the accuracy of three models of glucose meters. *Ann. Lab. Med.* 36, 271–274.
137. Ceriotti, F., Kaczmarek, E., Guerra, E., Mastrantonio, F., Lucarelli, F., Valgimigli, F. and Mosca, A. 2015. Comparative performance assessment of point-of-care testing devices for measuring glucose and ketones at the patient bedside. *J. Diabetes Sci. Technol.* 9, 268–277.
138. Sartor, Z., Kesey, J. and Dissanaike, S. 2015. The effects of intravenous vitamin C on point-of-care glucose monitoring. *J. Burn Care Res.* 36, 50–56.
139. Smith, K. E., Brown, C. S., Manning, B. M., May, T., Riker, R. R., Lerwick, P. A., Hayes, T. L. and Fraser, G. L. 2018. Accuracy of point-of-care blood glucose level measurements in critically ill patients with sepsis receiving high-dose intravenous vitamin C. *Pharmacotherapy* 38, 1155–1161.
140. Vasudevan, S. and Hirsch, I. B. 2014. Interference of intravenous vitamin C with blood glucose testing. *Diabetes Care* 37, e93–94.
141. Ma, Y., Sullivan, G. G., Schrick, E., Choi, I. Y., He, Z., Lierman, J., Lee, P., Drisko, J. A. and Chen, Q. 2013. A convenient method for measuring blood ascorbate concentrations in patients receiving high-dose intravenous ascorbate. *J. Am. Coll. Nutr.* 32, 187–193.
142. Carr, A. C., Vissers, M. C. and Cook, J. S. 2014. The effect of intravenous vitamin C on cancer- and chemotherapy-related fatigue and quality of life. *Front. Oncol.* 4, 283.
143. Carr, A. C. and McCall, C. 2017. The role of vitamin C in the treatment of pain: new insights. *J. Transl. Med.* 15, 77.
144. Du, J., Cullen, J. J. and Buettner, G. R. 2012. Ascorbic acid: Chemistry, biology and the treatment of cancer. *Biochim. Biophys. Acta.* 1826, 443–457.
145. Hoffer, L. J., Robitaille, L., Zakarian, R., Melnychuk, D., Kavan, P., Agulnik, J., Cohen, V., Small, D. and Miller, W. H., Jr. 2015. High-dose intravenous vitamin C combined with cytotoxic chemotherapy in patients with advanced cancer: A phase I-II clinical trial. *PLoS One* 10, e0120228.
146. Yang, H., Villani, R. M., Wang, H., Simpson, M. J., Roberts, M. S., Tang, M. and Liang, X. 2018. The role of cellular reactive oxygen species in cancer chemotherapy. *J. Exp. Clin. Cancer Res.* 37, 266.

147. Espey, M. G., Chen, Q. and Levine, M. 2009. Comment re: Vitamin C antagonizes the cytotoxic effects of chemotherapy. *Cancer Res.* 69, 8830; author reply 8830–8831.
148. Heaney, M. L., Gardner, J. R., Karasavvas, N., Golde, D. W., Scheinberg, D. A., Smith, E. A. and O'Connor, O. A. 2008. Vitamin C antagonizes the cytotoxic effects of antineoplastic drugs. *Cancer Res.* 68, 8031–8038.
149. Klimant, E., Wright, H., Rubin, D., Seely, D. and Markman, M. 2018. Intravenous vitamin C in the supportive care of cancer patients: A review and rational approach. *Curr. Oncol.* 25, 139–148.
150. Weijl, N. I., Hopman, G. D., Wipkink-Bakker, A., Lentjes, E. G., Berger, H. M., Cleton, F. J. and Osanto, S. 1998. Cisplatin combination chemotherapy induces a fall in plasma antioxidants of cancer patients. *Ann. Oncol.* 9, 1331–1337.
151. Jonas, C. R., Puckett, A. B., Jones, D. P., Griffith, D. P., Szeszycki, E. E., Bergman, G. F., Furr, C. E. et al. 2000. Plasma antioxidant status after high-dose chemotherapy: A randomized trial of parenteral nutrition in bone marrow transplantation patients. *Am. J. Clin. Nutr.* 72, 181–189.
152. Hunnisett, A., Davies, S., McLaren-Howard, J., Gravett, P., Finn, M. and Gueret-Wardle, D. 1995. Lipoperoxides as an index of free radical activity in bone marrow transplant recipients. Preliminary observations. *Biol. Trace Elem. Res.* 47, 125–132.
153. Goncalves, T. L., Benvegnu, D. M., Bonfanti, G., Frediani, A. V. and Rocha, J. B. 2009. delta-Aminolevulinate dehydratase activity and oxidative stress during melphalan and cyclophosphamide-BCNU-etoposide (CBV) conditioning regimens in autologous bone marrow transplantation patients. *Pharmacol. Res.* 59, 279–284.
154. Nannya, Y., Shinohara, A., Ichikawa, M. and Kurokawa, M. 2014. Serial profile of vitamins and trace elements during the acute phase of allogeneic stem cell transplantation. *Biol. Blood Marrow Transplant.* 20, 430–434.
155. Nielsen, T. K., Hojgaard, M., Andersen, J. T., Poulsen, H. E., Lykkesfeldt, J. and Mikines, K. J. 2015. Elimination of ascorbic acid after high-dose infusion in prostate cancer patients: A pharmacokinetic evaluation. *Basic Clin. Pharmacol. Toxicol.* 116, 343–348.
156. Seely, D., Stempak, D. and Baruchel, S. 2007. A strategy for controlling potential interactions between natural health products and chemotherapy: A review in pediatric oncology. *J. Pediatr. Hematol. Oncol.* 29, 32–47.
157. Grasso, C., Fabre, M. S., Collis, S. V., Castro, M. L., Field, C. S., Schleich, N., McConnell, M. J. and Herst, P. M. 2014. Pharmacological doses of daily ascorbate protect tumors from radiation damage after a single dose of radiation in an intracranial mouse glioma model. *Front. Oncol.* 4, 356.
158. Cullen, J. J. 2018. The dual effect of pharmacological ascorbate on radiation: The best of both worlds. *Oncotarget* 9, 36648–36649.
159. Park, J. H., Davis, K. R., Lee, G., Jung, M., Jung, Y., Park, J., Yi, S. Y. et al. 2012. Ascorbic acid alleviates toxicity of paclitaxel without interfering with the anticancer efficacy in mice. *Nutr. Res.* 32, 873–883.
160. Wang, C., Lv, H., Yang, W., Li, T., Fang, T., Lv, G., Han, Q. et al. 2017. SVCT-2 determines the sensitivity to ascorbate-induced cell death in cholangiocarcinoma cell lines and patient derived xenografts. *Cancer Lett.* 398, 1–11.
161. Xia, J., Xu, H., Zhang, X., Allamargot, C., Coleman, K. L., Nessler, R., Frech, I., Tricot, G. and Zhan, F. 2017. Multiple myeloma tumor cells are selectively killed by pharmacologically-dosed ascorbic acid. *EBioMedicine* 18, 41–49.
162. Serrano, O. K., Parrow, N. L., Violet, P. C., Yang, J., Zornjak, J., Basseville, A. and Levine, M. 2015. Antitumor effect of pharmacologic ascorbate in the B16 murine melanoma model. *Free Radic. Biol. Med.* 87, 193–203.
163. Ou, J., Zhu, X., Lu, Y., Zhao, C., Zhang, H., Wang, X., Gui, X. et al. 2017. The safety and pharmacokinetics of high dose intravenous ascorbic acid synergy with modulated electrohyperthermia in Chinese patients with stage III–IV non-small cell lung cancer. *Eur. J. Pharm. Sci.* 109, 412–418.
164. Bruckner, H., Hirschfeld, A., Gurell, D. and Lee, K. 2017. Broad safety impact of high-dose ascorbic acid and induction chemotherapy for high-risk pancreatic cancer. *J. Clin. Oncol.* 35(Suppl), e15711–e15711.
165. Du, J., Cieslak, J. A., 3rd, Welsh, J. L., Sibenaller, Z. A., Allen, B. G., Wagner, B. A., Kalen, A. L. et al. 2015. Pharmacological ascorbate radiosensitizes pancreatic cancer. *Cancer Res.* 75, 3314–3326.

166. Cieslak, J. A., Sibenaller, Z. A., Walsh, S. A., Ponto, L. L., Du, J., Sunderland, J. J. and Cullen, J. J. 2016. Fluorine-18-labeled thymidine positron emission tomography (FLT-PET) as an index of cell proliferation after pharmacological ascorbate-based therapy. *Radiat. Res.* 185, 31–38.
167. Vollbracht, C., Schneider, B., Leendert, V., Weiss, G., Auerbach, L. and Beuth, J. 2011. Intravenous vitamin C administration improves quality of life in breast cancer patients during chemo-/radiotherapy and aftercare: results of a retrospective, multicentre, epidemiological cohort study in Germany. *In Vivo.* 25, 983–990.
168. Ngo, B., Van Riper, J. M., Cantley, L. C. and Yun, J. 2019. Targeting cancer vulnerabilities with high-dose vitamin C. *Nat. Rev. Cancer* 19, 271–282.
169. Padayatty, S. J., Sun, A. Y., Chen, Q., Espey, M. G., Drisko, J. and Levine, M. 2010. Vitamin C: Intravenous use by complementary and alternative medicine practitioners and adverse effects. *PLoS One* 5, e11414.
170. Mustafi, S., Camarena, V., Volmar, C. H., Huff, T. C., Sant, D. W., Brothers, S. P., Liu, Z. J., Wahlestedt, C. and Wang, G. 2018. Vitamin C sensitizes melanoma to BET inhibitors. *Cancer Res.* 78, 572–583.
171. Kawada, H., Sawanobori, M., Tsuma-Kaneko, M., Wasada, I., Miyamoto, M., Murayama, H., Toyosaki, M. et al. 2014. Phase I clinical trial of intravenous L-ascorbic acid following salvage chemotherapy for relapsed B-cell non-Hodgkin's lymphoma. *Tokai J. Exp. Clin. Med.* 39, 111–115.
172. Riordan, H. D., Casciari, J. J., Gonzalez, M. J., Riordan, N. H., Miranda-Massari, J. R., Taylor, P. and Jackson, J. A. 2005. A pilot clinical study of continuous intravenous ascorbate in terminal cancer patients. *P. R. Health Sci. J.* 24, 269–276.
173. Shim, E. H., Livi, C. B., Rakheja, D., Tan, J., Benson, D., Parekh, V., Kho, E. Y. et al. 2014. L-2-Hydroxyglutarate: An epigenetic modifier and putative oncometabolite in renal cancer. *Cancer Discov.* 4, 1290–1298.
174. Uetaki, M., Tabata, S., Nakasuka, F., Soga, T. and Tomita, M. 2015. Metabolomic alterations in human cancer cells by vitamin C-induced oxidative stress. *Sci. Rep.* 5, 13896.

CHAPTER EIGHT

Orthomolecular Approaches for the Use of Intravenous Vitamin C

ADMINISTRATION BY INTEGRATIVE AND NATUROPATHIC DOCTORS

Jeanne A. Drisko

CONTENTS

INTRODUCTION

Intravenous vitamin C (IVC) history and use have taken a convoluted path to the present, setting out in the first half of the twentieth century with great promise for parenteral use of vitamin C as an antiviral and antibacterial agent with practical biochemical applications. However, entry into the antibiotic and vaccine era and a negative scientific report effectively stopped further research, driving the practice of parenteral vitamin C underground. Since the practice was outside the purview of conventional medicine, therapeutic vitamin C use was forgotten during the mid-portion of the twentieth century except by a small minority of nutritionally minded practitioners.

In 2010, our group published a report of a survey asking integrative physicians if they used IVC, and if so, how [1]. What was known prior to the survey was that integrative medicine physicians, including naturopathic doctors, were using IVC as part of their practice, but the scale of use and types of adverse events were unknown. Attendees at integrative medicine conferences in 2006 and 2008 were surveyed, and in parallel, sales of IVC by major U.S. manufacturers and distributors were determined. As part of the survey, information was collected from practitioners regarding adverse events identified. Published cases and the U.S. Food and Drug Administration's (FDA) Adverse Event Reporting System database were also analyzed for reports of adverse events specific to vitamin C. Of 199 surveys returned out of 550, 172 practitioners administered IVC to 11,233 patients in 2006 and 8876 patients in 2008.

By report, the average dose administered was 28 g given every 4 days, with 22 total treatments per patient. In the United States, most dosing vials contain 25 g in 50 mL. Estimated yearly numbers of vials used by the surveyed practitioners were

318,539 in 2006 and 354,647 in 2008. This contrasts with manufacturers' yearly sales of 750,000 and 855,000 vials, respectively, for the same reporting period.

Common reported reasons for treatment included infection, cancer, and fatigue. Of 9328 patients for whom data are available, 101 had side effects, mostly minor, including lethargy/fatigue in 59 patients, short-lived change in mental status in 21 patients, and vein irritation/phlebitis in 6 patients. Review of the published literature documented serious adverse events, including two deaths in patients known to be at risk for receiving IVC. Secondary to multiple prescribed medications and associated underlying chronic illness, the FDA Adverse Event Reporting System database was uninformative for specific information pinpointing adverse events directly related to IVC.

Accurate numbers of patients treated in the United States with high-dose parenteral vitamin C could not be estimated from the survey because those responding represented only a fraction of doctors administering IVC. But what is known is that high-dose IVC is in wide use by integrative practitioners [1]. Conventional physicians were made aware of this use by the survey publication and cautioned to inquire about IVC use in patients with cancer and other chronic, untreatable, or intractable conditions and to be observant of unexpected harm, drug interactions, or benefits.

As the survey suggested, and borne out by ongoing translational research, other than the known complications of IVC administration in those with renal impairment or glucose-6-phosphate dehydrogenase (G6PD) deficiency, high-dose IVC appears to be remarkably safe [2–9].

The use of pharmacologic IVC is not unknown to those practicing in orthomolecular, integrative, or naturopathic medical paradigms. The infusion of vitamins, minerals, amino acids, and other natural substances are tools used to treat patients. The practice of IVC, however, is unknown to most conventional medical doctors. How two very divergent paths came about is apparent only after understanding the circuitous route parenteral vitamin C has traveled.

HISTORY OF USE FROM AN ORTHOMOLECULAR WORLDVIEW

The birth of vitamin C sprang from the scientific efforts of Albert Szent-Györgyi (1893–1986) and others [10,11]. In 1907, Axel Holst and Alfred Fröhlich theorized the existence of vitamin C based on its biological effects, and an international competition was announced to spur isolation the vitamin. In 1928, Albert Szent-Györgyi isolated a substance identified 4 years later as vitamin C, which led to the 1937 Nobel Prize in Physiology or Medicine for the discovery. The subsequent work of Szent-Györgyi and other investigators finally explained the link between vitamin C and the treatment and prevention of scurvy.

Since the eighteenth century, it had been known that citrus fruit prevented scurvy, but no one knew what it was in the citrus that warded off the condition. Szent-Györgyi made the observation that some fruits turned brown when cut and exposed to air, while others, such as lemons and oranges, did not. Remarkably, he noticed a similarity between this color change and the bronzing of the human skin in people with Addison disease. It was his famous intellect that led to the link between vitamin C in foods and vitamin C in human biologic systems. Szent-Györgyi's credo was "to see what everyone else has seen, but think what no one else has thought," and he set off on a 10-year course of experimentation and investigation. Readers wanting more of the story are encouraged to read Ralph Moss's biography of Szent-Györgyi [10].

The early years of the twentieth century led to expansive research in all the vitamins, leading to numerous Nobel Prize awards and ways to introduce clinical applications [11]. Expanding interest in the use of vitamin therapies began in earnest.

STANDING ON THE SHOULDERS OF GIANTS

Claus Washington Jungeblut, MD

Claus Washington Jungeblut is considered the father of clinical applications of IVC, although his contributions are largely lost in the shadows of time [12–17]. Jungeblut received his doctor of medicine degree from the University of Bern, Switzerland, in 1921, and between 1921 and 1923, he conducted research at the Robert Koch Institute in Berlin [18]. He moved to New York and became a bacteriologist for the New York State Department of Health from 1923 until 1927. Subsequently, Jungeblut became associate professor at Stanford University from 1927 until 1929. Thereafter, he joined the faculty at the Columbia University College of Physicians and Surgeons in New York City as a professor of bacteriology and retired on June 30, 1962 [18].

The *New York Times* obituary archives acknowledge his role in poliomyelitis research by reporting, "Dr. Claus W. Jungeblut, former professor of bacteriology at Columbia University, who was well known for his research in infantile paralysis, died yesterday at his home in Westport, Conn. He was 78 years old. In 1942 Dr. Jungeblut and Dr. Murray Sanders reported on transferring a polio virus from monkeys to rats to mice and developing a changed virus that protected monkeys and prevented their paralysis if used in time" [18]. His rich preclinical research and publication record in parenteral vitamin C are not noted, although a record of some of his publications on IVC exist [12,17,19].

One of Jungeblut's earliest research findings was ascorbic acid's ability to neutralize and render harmless many bacterial toxins, such as tetanus, diphtheria, and *Staphylococcus* toxins [13,15,20]. Jungeblut's reports gave rise to other investigators replicating his findings using vitamin C in a variety of bacterial and viral diseases [21]. Between 1936 and 1937, scientists demonstrated similar inactivation of infectious agents: Holden et al. using herpes virus; Kligler and Bernkopf in vaccinia virus; Lagenbusch and Enderling with the virus of hoof-and-mouth disease; Amato in rabies virus; Lominski using bacteriophage; and Lojkin and Martin with the tobacco mosaic disease virus [21]. Even at this early date, it was established that ascorbic acid had the potential of being a wide-spectrum antiviral agent [22]. But like so many medical visionaries before, Jungeblut's discoveries were countered and ignored.

The blow dealt to Jungeblut's vitamin C research in treating multiple infectious illnesses, especially poliomyelitis, came at the hands of Albert B. Sabin of the famous Sabin live polio vaccine. Sabin, attempting to reproduce Jungeblut's work with vitamin C treatment in polio-infected monkeys, failed to obtain any positive results in inhibiting the onset of polio infection [23]. Both scientists became bogged down arguing the technical details of the experiments with each other. Presaging other vitamin C experiments to come decades later, it is now apparent that the parenteral high-dosing method used by Jungeblut was not adhered to by Sabin, whose report focused on low, oral dosing. The amount of vitamin C and the frequency of the dosing used by Sabin were insufficient to maintain high levels of ascorbic acid in the blood during the incubation of the disease [20,22,23]. Unfortunately, Sabin's publication reporting the negative value of vitamin C in the treatment of infectious disease was accepted as final. The result of the negative report had a chilling effect on further vitamin C research applications. This effectively relegated parenteral vitamin C therapy to a small cadre of practitioners from the 1940s onward, the true numbers of which we will never know [20,22].

Frederick Klenner, MD

In Reidsville, North Carolina, a small-town physician, Frederick R. Klenner, was one of the practitioners who carried on the use of IV and intramuscular (IM) injections of nutrients, especially employing vitamin C for many acute and chronic illnesses. Klenner is likely remembered because he published articles detailing his practice and outcomes, unlike a cadre of other unpublished practitioners from that time period, whose practices of orthomolecular medicine are lost [24–29]. Klenner was helpful in keeping the orthomolecular practice of high-dose vitamins and minerals alive for those who followed.

Born in 1907, Klenner earned his undergraduate and graduate degrees in biology, magna cum laude, from St. Vincent and St. Francis Colleges [27]. Klenner graduated from Duke University School of Medicine in 1936 and subsequently completed his residency at the North Carolina Pickford Tuberculosis Sanitarium. After his postgraduate hospital training, Klenner entered private practice in medicine. Although specializing in diseases of the chest, he "continued to do general practice because of the opportunities it afforded for observations in medicine" [27].

Klenner's entry into the use of high-dose parenteral nutrients began in 1942 when he successfully treated his wife's chronic periodontal disease with parenteral vitamin C, halting the need for full dental extractions [28]. This led to his use of IM vitamin C to successfully treat a moribund patient with viral pneumonia [26]. After these events, his use of oral, IM, and IV high-dose vitamins began in earnest, and Klenner documented a wide variety of illnesses treated [24–29]. Klenner subsequently became a student of Jungeblut and others' scholarly works from the earlier period before vitamin therapies fell out of favor [28]. Noting the importance of the early research and the lack of interest in his day, Klenner wrote, "Some physicians would stand by and see their patient die rather than use ascorbic acid because in their finite minds it exists only as a vitamin" [29].

Klenner commented on Jungeblut's earlier work, stating that some reported results were inconclusive because the amount of vitamin C given by Jungeblut was inadequate to cope with the degree of infection [26], but he commented that Jungeblut's practice of giving higher doses by "needle" provided adequate dosing in some experiments. Klenner further commented that Sabin's results were negative compared to Jungeblut's because Sabin used a greater dose of virus and less vitamin C. Klenner concluded that if high blood and tissue levels of ascorbic acid are continuously maintained, "an extremely unfavorable environment for viral growth and reproduction is created in the human body" [26]. In summary, Klenner's maxim was that the degree of neutralization of infectious disease was always in proportion to the concentration of administered vitamin C and the length of time it was employed.

Klenner administered ascorbate by injection as outlined in his publications, stating the most effective route was intravenously, but the IM route was also effective [24–27]. He administered at least 350 mg per kilogram of body weight per day, approximating a dose of 25–30 g for an adult. Klenner reported that 350 mg per kilogram of body weight in divided doses every 2 hours would effectively "stop measles and dry up chicken pox" [26,27]. Such use exemplifies the modern orthomolecular physician with dosing amounts and schedules flexible and symptom driven [28]. The sicker the patient, the higher is the dose. Massive ascorbate treatment cured every one of 60 polio cases Klenner saw, by his report [24]. All patients were well within the week, and none progressed to paralysis according to his case reports.

Other case reports of uses of parenteral vitamins and minerals, particularly vitamin C, document myriad conditions treated by these approaches [27]. Unfortunately, at Klenner's death in 1984, his office was closed, and his medical records and other papers were burned, leaving a hole in the art and practice of orthomolecular medicine* [30].

* The dark history of the Klenner family leading to the clinic closure and the destruction of medical records can be better understood by reading J. Bledsoe, *Bitter Blood: A Story of Southern Family Pride, Madness, and Multiple Murder* (New York, NY: EP Dutton/Penguin, 1988).

Mark Levine, MD

Because the details of Levine's research are well detailed in this textbook in other chapters, they are not repeated here. It would be remiss if not noted that Levine's ascorbic acid research has provided a cornerstone on which to build further translational research [31–35].

Hugh D. Riordan, MD, Orthomolecular Pioneer

The word *orthomolecular* was created by Linus Pauling, who is famously known for his contributions to the field of biochemistry, creating the underpinning for much of modern medicine. Pauling's orthomolecular views on the use of megadoses of vitamins became unpopular in mainstream medicine, yet many pioneering physicians advocated orthomolecular treatments.

Conventional medicine continues to adhere to the belief that doses above the recommended daily allowances may be harmful at worst, and unnecessary at the very least, while continuing to focus on deficiency diseases as the only use for vitamins in the treatment of patients. Despite opposition from conventional medical paradigms, orthomolecular practices continued to grow, and one pioneer who made that possible was Hugh D. Riordan.

Abram Hoffer wrote a tribute to Riordan after his death on January 7, 2005 [36]. In the tribute, Hoffer noted orthomolecular medicine emphasizes the importance of nutrition and also augmenting the diet with high-dose nutrients given orally or parenterally for those in need. Those in need may have biochemical abnormalities, poor diets, or illnesses that even the best of diets will be insufficient to support. Hoffer pointed to Riordan as a leader of orthomolecular medicine.

Riordan established the Center for the Improvement of Human Functioning, now called the Riordan Center, where he practiced orthomolecular medicine. One of his myriad interests was the use of IVC in high doses for the treatment of cancer. After the 1996 publication of Mark Levine [31], Riordan sought out Levine to share the results of cancer patients treated at the center using high-dose IVC. This meeting led to a National Institutes of Health (NIH) systematic review of three patients treated at the center, and a publication of the findings followed [37]. Of interest, one of the coauthors, John Hoffer, a well-regarded clinician at McGill University in

Montréal, Canada, is the son of Abram Hoffer. Riordan was one of the first to suggest how large doses of vitamin C could be chemotherapeutic for cancer patients.

Hoffer's tribute to Riordan covered a wide range of Riordan's uses of orthomolecular medicine and rightly stated that because of his practice, "he became a member of an elite group that includes Linus Pauling, Roger Williams, Carl Pfeiffer, Humphry Osmond, Irwin Stone," and others, including Abram Hoffer himself [36]. This chapter's author was fortunate to have been trained by Riordan in the use of orthomolecular medicine and carry on research in IVC.

John Myers, MD

One of the unknown orthomolecular physicians who practiced in relative obscurity in Maryland was Myers. Myers would have remained unknown to many if it were not for Alan Gaby's paper on the use of the Myers' Cocktail—a mixture of high-dose vitamin C and other vitamins and minerals given by IV push via syringe [38]. Gaby named the infusion after Myers, and it famously bears his name today, although little is known of the physician himself. How remarkable it is that across the United States there were likely many physicians who used large doses of vitamins and minerals in the treatment of illnesses who have long been forgotten and who remain unsung heroes [30].

Other Orthomolecular Proponents of Note

Pioneers of orthomolecular medicine include Linus Pauling, Ewan Cameron, Lendon Smith, Irwin Stone, Abram Hoffer, Robert Cathcart, and those unknown courageous practitioners of orthomolecular medicine who toiled in obscurity for the betterment of humankind [39].

DELIVERY OF ORTHOMOLECULAR VITAMIN C IN THE CLINICAL SETTING

Through the legacy of pioneers of orthomolecular medicine, many practitioners continue to treat patients with parenterally infused vitamin C. Clinical uses of high-dose parenteral vitamin C include cardiovascular disease (with vitamin C as part of the infusate) [40–43], chronic fatigue [44,45], infections [1,38] including sepsis [46–50], diabetes [41,51–54], cancer [4–7,9,33–35,55–59], pain [60,61], and other uses [32,62,63]. For the purposes of this chapter, the focus is on the treatment of cancer, although general information about infusion preparation is included.

Well known is the use of IVC as a chemotherapeutic agent beginning with the collaboration between Linus Pauling and Ewan Cameron and continuing to the current time [5–7,37,55,64,65]. This is despite the controversy that arose because of the Mayo Clinic trials documented elsewhere in this textbook [66]. Case reports continue to show benefits that should be compelling to shape research agendas [37,59,67,68], and phase I and small phase II trials using IVC in a variety of cancers, largely funded by private foundations, continue to be published [5–9]. Yet funding of larger trials at the federal level continues to be fraught with controversy because of ongoing misinformation and misunderstanding perpetuated since the outdated Mayo Clinic trials were reported. The way forward is to continue translational research that is revealing mechanisms of action and biological pathways [4–7,9,55,69,70].

Notwithstanding difficulties and controversies facing the pharmacologic ascorbate research agenda, research and use as a cancer therapy continues around the world [1,4,71–77]. The preparation, administration, and dosing of IVC varies, but the practice of IV administration is a constant. It is now known from Levine and colleagues [31,33–35] that only parenteral administration of vitamin C bypasses tight control of plasma and tissue levels that are imposed by bowel absorption and kidney excretion as contrasted to vitamin C given orally. This is a fundamental tenet of IVC administration and leads to the understanding that IVC is a pharmacologic agent, no longer acting as a vitamin as it exists at lower plasma concentrations. No amount or type of orally administered vitamin C will attain plasma levels reached by parenteral administration.

Another fundamental tenet is the pharmacologic action of IVC to produce hydrogen peroxide and act as a pro-oxidant (see other chapters in this text for more in-depth explanation) [8,33–35,55,69,78–80], again in contrast to its antioxidant properties when vitamin C is administered orally. With these understandings and augmented by new mechanistic pathways, we are learning how IVC acts as a chemotherapeutic agent and shows synergism with conventional cancer therapy [4–7,9].

What may be lost in current scientific reports are nuances of in-office administration that may be appreciated when patients are treated outside of protocols. Those of us who trained with the pioneers learned by sitting with patients and observing their reactions during IV administration, documenting progress or lack thereof through laboratory testing and imaging, and learning to make decisions about dosing. After all, this is about patients, not about paradigms.

IVC does not act rapidly when given as a chemotherapeutic agent [6,7,59,67]. Therefore, patients at the end of their lives and heavily pretreated with chemotherapy and/or radiation therapy rarely benefit from IVC. Rather, in therapy, IVC is to be used early in the process of cancer treatment and can be considered an adjunctive therapy when chemotherapy and/or radiation therapy are considered [4–7,9].

Because it appears to act slowly, several months may pass before benefit is seen through reduction in tumor markers or by imaging [67]. Giving several doses sporadically in 1 month's time is not a fair trial to determine patients' responses to IVC. Rather, consistent administration two to three times per week, at a minimum for 2–3 months, is required before repeat imaging and tumor markers are reevaluated to determine response [5–7]. Also, as understood by our naturopathic colleagues, the terrain must be prepared to fully accept lifesaving therapies during cancer treatment [81,82]. Then when treated by radiation and/or chemotherapy with the expected common side effects of appetite reduction and nausea and vomiting, the patients are better supported. When combining IVC with cancer therapies, an additional effect of mood elevation may occur [6,83].

Paramount in the proper administration of IVC is the concept of dose escalation. Beginning with a lower dose, for example a 25 g dose, initial administration is given to assess patients' tolerability to the infusion and to prepare for the experience. The dose can be escalated to 50 g, for example, at the next infusion, but the critical step is to evaluate the patient's plasma vitamin C level during dose escalation. Anecdotal reports from practitioners seem to point to 350–450 mg/dL as a target plasma dose, and this corresponds to preclinical research findings that describe ranges around 20 mM concentrations affecting neoplastic cell kill [33,35,62,84,85]. Individual patients with different tumor types have varying requirements in the amount of vitamin C dosing necessary to hit a target plasma level. It appears the more aggressive the tumor type, the higher the dose of vitamin C is required to reach targeted plasma levels. Rather than relying on milligram/kilogram dosing schedules, it is more productive to follow the plasma vitamin C level. We know that plasma concentrations in millimole per deciliter (mM/dL) ranges translated to milligram per deciliter (mg/dL) can advise the amount of vitamin C administered, guiding targeted plasma level ranges that may be therapeutic [6,7].

The caveat is that plasma vitamin C levels are difficult to measure because of the rapid oxidation and disappearance of vitamin C in the plasma and blood specimen. During dose escalation, baseline blood is drawn prior to the start of the infusion. Once the infusion begins, the practitioner has to be prepared to redraw the blood immediately at the termination of the infusion. The patient cannot be sent to the lab for the blood draw, as the plasma vitamin C concentration will rapidly fall. If the blood is drawn in the office at the conclusion of the infusion, it cannot sit on the counter waiting to be sent to the lab at a later time, as the vitamin C will rapidly degrade in the specimen. With the vitamin C rapidly metabolizing in the blood specimen, if proper sample acquisition and preparation are not adhered to, accurate plasma vitamin C levels cannot be determined.

Immediately after the blood sample is collected, it must be covered from light, placed on ice, and taken to the lab to be handled as a critical frozen specimen and should be assayed within minutes. This is unlikely to occur in the majority of clinical settings, even those with an on-site laboratory. We were fortunate as a translational research team to be able to draw the specimen, place it on ice, and transfer it immediately to our research lab where it was assayed by high-performance liquid chromatography (HPLC).

Because these circumstances do not exist for most practitioners, this prevents many from obtaining an accurate vitamin C plasma reading. As a result of this difficulty, it is possible to exploit a well-known lab error that occurs when patients are receiving IVC. In brief, while obtaining fingerstick glucometer readings, the molecule of vitamin C is read as a molecule of glucose because of their similar molecular structures [86]. If a patient receives an infusion of vitamin C and immediately checks glucose levels by fingerstick glucometer, it will read as elevated glucose. But the glucometer has given this reading in error because the elevated plasma vitamin C has been read as glucose by the glucometer. Therefore, a baseline glucometer reading is obtained, the IVC infusion given, and immediately at the end of the infusion a repeat glucometer reading is obtained. Once the two values are subtracted, the remaining value represents a ballpark estimate of the plasma vitamin C level in milligrams per deciliter. That is, the baseline reading represents the glucose level, while the postinfusion reading represents the glucose level plus the vitamin C level together, and subtracting will give you the estimated vitamin C level. This simple solution has been validated and proven helpful [86].

A warning to all—do not treat the post-IVC fingerstick glucometer reading as elevated glucose to be treated by insulin injection. This will result in hypoglycemia. If a concern for elevated glucose exists around the time of IVC infusion, the patient must have a conventional blood draw in the lab to get an accurate glucose level. The conventional blood test in the lab will not confuse vitamin C with glucose as occurs with the fingerstick glucometer method because of differences in the assay method.

If the plasma vitamin C level is not in an adequate range of 350–450 mg/dL, then further dose escalation is necessary. For example, after administration of 50 g, if the plasma level is only 280 mg/dL, this would not be considered in a therapeutic range. The dose would need to be increased to 75 g of vitamin C. It is convenient to dose by increasing by 25 g increments because most injectable vitamin C in the United States is available in 25 g bottles usually as 50 mL volume. Any dose of a drug is designed to produce a plasma, and hopefully site of action, concentration most likely to produce the desired therapeutic effect and to minimize adverse effect. This is determined by several factors, but perhaps the most relevant is the volume of distribution. Volume of distribution, like blood or plasma volume, organ size, and other anatomic/pharmacologic factors, is in some way proportional to body mass or size. A standard adult or pediatric dose is based on the average-sized child or adult. With drugs that have wide therapeutic windows, the use of this average value is safe. For drugs with a very narrow therapeutic index (cardiac glycosides, warfarin, cytotoxic chemotherapeutic drugs, etc.), one size definitely does not fit all. These are the cases where dose normalization is most commonly employed. Normalization by weight (mg/kg) or body surface area may be used; both are attempts to scale the dose to the volume of distribution, resulting in a closer to ideal drug concentration in the patient. Changes in volume of distribution with body mass/size will directly affect Cmax, Css, and area under the curve (AUC) for a given dose of a drug, with all being inversely proportional to volume of distribution [87,88]. However, vitamin C is almost entirely cleared by renal excretion rather than CYP 450 hepatic clearance, and this occurs rapidly over several hours [32,89]. It is satisfactory to dose by incremental escalation and rely on plasma vitamin C levels to guide dosing.

Osmolality is another important concern in choosing the carrier fluid and volume for preparing the vitamin C infusion (Table 8.1). A hypotonic solution should never be infused, as it will result in red blood cell hemolysis, and this is basic infusion therapy safety for any type of infusion [87]. Of note, chemotherapeutic IVC administration in our clinical trials used additional magnesium with the IVC because magnesium minimizes venous spasm and irritation. Many other practitioners in private practices outside of research protocols add a variety of other nutrients, including minerals, vitamins, and botanicals. No information is currently available to determine if these are beneficial or not. At this time, historical use and reports of safety guide the practitioner's decision in the administration of parenteral nutrients. This has led to different groups proposing different methods that border on dogma without proof. Only future research can determine the correct approach and benefit.

In our translational research protocols, infusion rate is timed so that 0.5 g is infused per minute [6,7]. Therefore, 25 g is infused over 50 minutes, for example. However, in clinical practice, there are many infusion methods for vitamin C, including IV push via syringe [38], IVC by gravity or pump, and IM administration. No one method will be advocated here, as the clinical situation dictates the choice. However, since oncology treatment usually requires 50 g or higher doses of vitamin C, it is preferable that either gravity or pump infusion be used rather than IV push or IM for patient safety and comfort.

It is often inconceivable to the uninitiated practitioner that gram doses are infused at each setting, even more than 100 g. Yet, it has been known for decades that the administration of high-dose vitamin C is safe [24–26,28,62,64], and this has been confirmed in more recent research

TABLE 8.1

Osmolarity of ascorbic acid and magnesium chloride in different volumes and types of carrier fluids

Sodium Ascorbic Acid (calculated using 500 mg/mL ascorbic acid)	Osmolarity Calculated in Sterile Water			Osmolarity Calculated in Ringer Lactate		
	250 mL	500 mL	1000 mL	250 mL	500 mL	1000 mL
1 gram	72	36	18	**347**	**312**	**297**
5 grams	261	131	65	**528**	**404**	**342**
10 grams	**499**	249	125	**754**	**517**	**399**
15 grams	**737**	**368**	184	**981**	**630**	**455**
25 grams	1212	**606**	**303**	1433	**857**	**568**
30 grams	1449	**725**	**362**	1660	**970**	**625**
50 grams	2400	1200	**600**	2565	1423	**851**
60 grams	2875	1437	**719**	3018	1649	**965**
75 grams	3588	1794	**897**	3697	1989	1134
100 grams	4776	2388	**1194**	4829	2555	1427

NOTE: Osmolarity is calculated by withdrawing the equivalent volumes of ascorbic acid and magnesium chloride (MgCl) from the carrier fluid bag prior to injecting the ascorbic acid and MgCl into the bag, i.e., for 1 g (2 mL) of ascorbic acid and 2 mL MgCl, 4 mL is withdrawn from a 250 mL bag of Ringer lactate first. (Courtesy of Jeanne A. Drisko, MD, CNS, FACN.)

[4–7,9]. Oncologists often express concern about using vitamin C with chemotherapy because of the mistaken belief that vitamin C will act as an antioxidant. Research has confirmed that IVC, contrasted with oral administration, becomes a pro-oxidative therapy [8,32–35,55,69]. In fact, there may be an additive or synergistic effect when IVC is given with chemotherapy [6,7].

Administration of IVC has been given preceding chemotherapy administration on the same day, acting as the fluid loading dose [6], or has been given on alternative dates from chemotherapy [4,5,7,9]. Some practitioners report giving lower doses of chemotherapy when combined with IVC [76]. The Iowa group has shown that IVC can potentiate radiation therapy, again because of the pro-oxidative benefits of IVC [8]. Yet much more preclinical and translational research needs to proceed, and increased federal funding is mandatory to support this agenda.

CONCLUSION

Historically, IVC use in research has been remarkably controversial. In Jungeblut's era, further research was shut down by a published report that used insufficient doses and oral administration. Although research was halted, the therapy did not disappear entirely, as practitioners during the following decades used IM and IV vitamin C as an antiviral, antibiotic, and general tonic. It was adopted as an orthomolecular therapy by Linus Pauling, and he advocated for its use in cancer treatment after collaborating with Ewan Cameron. But once again, incorrect research approaches using insufficient doses with oral administration of vitamin C yielded negative outcomes, and the use of IVC was dismissed as a therapy from the late 1970s forward. These errors in research design came to light with the insights provided by Levine and colleagues contrasting oral dosing with parental administration. Parenteral dosing of IVC is not dependent on gut absorption and therefore is able to reach high plasma levels, bypassing tight control. Parenteral vitamin C administration is considered a drug and has been referred to as pharmacologic ascorbate, while oral vitamin C is under tight control and is considered a vitamin.

There continue to be practitioners around the world giving IV and IM vitamins, including vitamin C in a variety of doses, combined with the myriad of other molecules, and given rapidly or slowly for any number of conditions. Patients continue to seek out these therapies, and the connectivity of modern populations suggests this will not be halted anytime soon. Yet there is a paucity of research to determine if benefit exists. Because of this use and in consideration of patients, federal funders should cast aside their dated prejudices from prior poorly designed research and look with fresh eyes at the more recent translational research pointing to a new direction.

REFERENCES

1. Padayatty, S. J., Sun, A. Y., Chen, Q., Espey, M. G., Drisko, J. and Levine, M. 2010 Jan Vitamin C: Intravenous use by complementary and alternative medicine practitioners and adverse effects. *PLOS ONE* 5(7), e11414. Available from: http://www.pubmedcentral.nih.gov/articlerender.fcgi?artid=2898816&tool=pmcentrez&rendertype=abstract
2. Levine, M., Espey, M. G. and Chen, Q. Losing and finding a way at C: New promise for pharmacologic ascorbate in cancer treatment. *Free Radic. Biol. Med.* 2009 Jul 1 [cited 2012 Aug 21]. 47(1), 27–9. Available from: http://www.pubmedcentral.nih.gov/articlerender.fcgi?artid=2981594&tool=pmcentrez&rendertype=abstract
3. Levine, M., Padayatty, S. J. and Espey, M. G. 2011 Mar Vitamin C: A concentration-function approach yields pharmacology and therapeutic discoveries. *Adv. Nutr.* 2(2), 78–88. Available from: http://advances.nutrition.org/content/2/2/78.short
4. Hoffer, L. J., Robitaille, L., Zakarian, R., Melnychuk, D., Kavan, P., Agulnik, J. et al. 2015 High-dose intravenous vitamin C combined with cytotoxic chemotherapy in patients with advanced cancer: A phase I-II clinical trial. *PLOS ONE* 10(4), e0120228. Available from: http://dx.plos.org/10.1371/journal.pone.0120228
5. Monti, D. A., Mitchell, E., Bazzan, A. J., Littman, S., Zabrecky, G., Yeo, C. J. et al. 2012 Jan Phase I evaluation of intravenous ascorbic acid in combination with gemcitabine and erlotinib in patients with metastatic pancreatic cancer. *PLOS ONE* 7(1), e29794. Available from: http://www.pubmedcentral.nih.gov/articlerender.fcgi?artid=3260161&tool=pmcentrez&rendertype=abstract
6. Ma, Y., Chapman, J., Levine, M., Polireddy, K., Drisko, J. and Chen, Q. 2014 Feb 5 High-dose parenteral ascorbate enhanced chemosensitivity of ovarian cancer and reduced toxicity of chemotherapy. *Sci. Transl. Med.* 6(222), 222ra18. Available from: http://www.ncbi.nlm.nih.gov/pubmed/24500406
7. Polireddy, K., Dong, R., Reed, G., Yu, J., Chen, P., Williamson, S. et al. 2017 High dose parenteral ascorbate inhibited pancreatic cancer growth and metastasis: Mechanisms and a Phase I/IIa study. *Sci. Rep.* 7(1), 17188. Available from: http://www.nature.com/articles/s41598-017-17568-8
8. Moser, J. C., Rawal, M., Wagner, B. A., Du, J., Cullen, J. J. and Buettner, G. R. 2014 Pharmacological ascorbate and ionizing radiation (IR) increase labile iron in pancreatic cancer. *Redox. Biol.* 2, 22–7. Available from: http://dx.doi.org/10.1016/j.redox.2013.11.005
9. Welsh, J. L., Wagner, B. A., van't Erve, T. J., Zehr, P. S., Berg, D. J., Halfdanarson, T. R. et al. 2013 Mar Pharmacological ascorbate with gemcitabine for the control of metastatic and node-positive pancreatic cancer (PACMAN): Results from a phase I clinical trial. *Cancer Chemother. Pharmacol.* 71(3), 765–75. Available from: http://www.pubmedcentral.nih.gov/articlerender.fcgi?artid=3587047&tool=pmcentrez&rendertype=abstract
10. Moss, R. 1988 *Free Radical: Albert Szent-Györgi and the Battle Over Vitamin C*. Paragon House. 316 p.
11. Souganidis, E. 2012 Nobel laureates in the history of the vitamins. *Ann. Nutr. Metab.* 61(3), 265–9.
12. Jungeblut, C. W. 1935 Sep 30 Inactivation of poliomyelitis virus in vitro by crystalline vitamin C (ASCORBIC ACID). *J. Exp. Med.* 62(4), 517–21. Available from: http://www.ncbi.nlm.nih.gov/pubmed/19870431
13. Jungeblut, C. W. and Zwemer, R. L. 1935 May 1 Inactivation of diphtheria toxin in vivo and in vitro by crystalline Vitamin C (ascorbic acid). *Exp. Biol. Med.* 32(8), 1229–34. Available from: http://ebm.sagepub.com/lookup/doi/10.3181/00379727-32-8039C
14. Jungeblut, C. W. 1937 Jan 1 Vitamin C therapy and prophylaxis in experimental poliomyelitis. *J Exp Med.* 65(1), 127–46. Available from: http://www.jem.org/cgi/doi/10.1084/jem.65.1.127
15. Jungeblut, C. 1937 Inactivation of tetanus toxin by crystalline vitamin C (l-ascorbic acid). *J. Immunol.* 33, 203–14. Available from: https://www.cabdirect.org/cabdirect/abstract/19371403240
16. Saul, A. W. 2006 Claus Washington Jungeblut, M.D. polio pioneer: Ascorbate advocate. *J. Orthomol. Med.* 21(2), 102–6.
17. Jungeblut, C. W. A further contribution to vitamin C therapy in experimental poliomyelitis. *J. Exp. Med.* 1939. 70, 315–32.
18. Claus Washington Jungeblut Obituary. 1976 Feb 2 *New York Times* [Internet]. Available from: https://www.nytimes.com/1976/02/02/archives/claus-jungeblut-bacteriologist-78.html
19. Jungeblut, C. W. 1937 Oct 1 Further observations on vitamin C therapy in experimental poliomyelitis. *J. Exp. Med.* [Internet]. 66(4),

459–77. Available from: http://www.jem.org/cgi/doi/10.1084/jem.66.4.459
20. Landwehr, R. 1991 The origin of the 61-year stonewall of vitamin C. *J. Orthomol. Med.* (6), 99–103.
21. Stone, I. 1984 Fifty years of research on ascorbate and the genetics of scurvy: From a better flavored beer to homo sapiens ascorbicus. *J. Orthomol. Psychiatry.* 13(4), 280–4.
22. Stone, I. 1972 *The Healing Factor: Vitamin C against Disease* [Internet]. The Putnam Publishing Group New York. Available from: http://vitamincfoundation.org/stone/ch12-16/chap12-16.htm#C14
23. Sabin, A. B. 1938 Vitamin C in relation to experimental poliomyelitis: With incidental observations on certain manifestations in macacus rhesus monkeys on a scorbutic diet. *J. Exp. Med.* 69(4), 507–16.
24. Klenner, F. R. 1949 Jul The treatment of poliomyelitis and other virus diseases with vitamin C. *South Med. Surg.* 111(7), 209–14. Available from: http://www.ncbi.nlm.nih.gov/pubmed/18147027
25. Klenner, F. R. 1951 Apr Massive doses of vitamin C and the virus diseases. *South. Med. Surg.* 113(4), 101–7. Available from: http://www.ncbi.nlm.nih.gov/pubmed/14855098
26. Klenner, F. R. 1953 The use of vitamin C as an antibiotic. *J. Appl. Nutr. [Internet]* 6(May), 274–8. Available from: https://www.seanet.com/~alexs/ascorbate/195x/klenner-fr-j_appl_nutr-1953-v6-p274.htm
27. Klenner, F. R. 1971 Observations on the dose and administration of ascorbic acid when employed beyond the range of a vitamin in human pathology. *J. Appl. Nutr.* 23(3–4), 61–8. Available from: http://orthomolecular.org/library/jom/1998/pdf/1998-v13n04-p198.pdf
28. Smith, L.H. 1988 Clinical Guide to the Use of Vitamin C: The Clinical Experiences of Frederick R. Klenner, M.D. abbreviated, sumarized and annotated by Lendon H. Smith, MD. Available from: https://www.seanet.com/~alexs/ascorbate/198x/smith-lh-clinical_guide_1988.htm
29. Saul, A. W. 2007 Hidden in plain sight: The pioneering work of Frederick Robert Klenner, M.D. *J. Orthomol. Med.* 22(1), 31–8.
30. McCracken, R. D. 2009 *Injectable Vitamin C and the Treatment of Viral and Other Diseases*, 2nd ed. (McCracken, R. D., ed), Hygea Publishing Company, Long Beach, CA.
31. Levine, M., Conry-Cantilena, C., Wang, Y., Welch, R. W., Washko, P. W., Dhariwal, K. R. et al. 1996 Apr 16 Vitamin C pharmacokinetics in healthy volunteers: Evidence for a recommended dietary allowance. *Proc. Natl. Acad. Sci. USA* 93(8), 3704–9. Available from: http://www.pubmedcentral.nih.gov/articlerender.fcgi?artid=39676&tool=pmcentrez&rendertype=abstract
32. Padayatty, S. J., Sun, H., Wang, Y., Riordan, H. D., Hewitt, S. M., Katz, A. et al. 2004 Apr 6 Vitamin C pharmacokinetics: Implications for oral and intravenous use. *Ann. Intern. Med.* 140(7), 533–7. Available from: http://www.desireerover.nl/wp-content/uploads/2011/10/Injectable-Vitamin-C.pdf#page=190
33. Chen, Q., Espey, M. G., Krishna, M. C., Mitchell, J. B., Corpe, C. P., Buettner, G. R. et al. 2005 Sep 20 Pharmacologic ascorbic acid concentrations selectively kill cancer cells: Action as a pro-drug to deliver hydrogen peroxide to tissues. *Proc. Natl. Acad. Sci. USA* 102(38), 13604–9. Available from: http://www.pubmedcentral.nih.gov/articlerender.fcgi?artid=1224653&tool=pmcentrez&rendertype=abstract
34. Chen, Q., Espey, M. G., Sun, A. Y., Lee, J.-H., Krishna, M. C., Shacter, E. et al. 2007 May 22 Ascorbate in pharmacologic concentrations selectively generates ascorbate radical and hydrogen peroxide in extracellular fluid in vivo. *Proc. Natl. Acad. Sci. USA* 104(21), 8749–54. Available from: http://www.pubmedcentral.nih.gov/articlerender.fcgi?artid=1885574&tool=pmcentrez&rendertype=abstract
35. Chen, Q., Espey, M. G., Sun, A. Y., Pooput, C., Kirk, K. L., Krishna, M. C. et al. 2008 Aug 12 Pharmacologic doses of ascorbate act as a prooxidant and decrease growth of aggressive tumor xenografts in mice. *Proc. Natl. Acad. Sci. USA* 105(32), 11105–9. Available from: http://www.pubmedcentral.nih.gov/articlerender.fcgi?artid=2516281&tool=pmcentrez&rendertype=abstract
36. Hoffer, A. 2005 A Tribute to Hugh D. Riordan, M.D., 1932 2005. *J. Orthomol. Med.* (C), 1–11. Available from: http://orthomolecular.org/history/hdrtribute.pdf
37. Padayatty, S. J., Riordan, H. D., Hewitt, S. M., Katz, A., Hoffer, L. J. and Levine, M. 2006 Mar 28 Intravenously administered vitamin C as cancer therapy: Three cases. *CMAJ.* 174(7), 937–42. Available from: http://www.pubmedcentral.nih.gov/articlerender.fcgi?artid=1405876&tool=pmcentrez&rendertype=abstract

38. G, A. R. 2002 Intravenous nutrient therapy: The "Myers" cocktail'. *Altern. Med. Rev.* 7(5), 389–403. Available from: http://uml.idm.oclc.org/login?url=http://search.ebscohost.com/login.aspx?direct=true&db=c8h&AN=106816993&site=ehost-live
39. Saul AW. http://orthomolecular.org/hof/index.shtml. Available from: http://orthomolecular.org/hof/index.shtml
40. Lamas, G. A., Goertz, C., Boineau, R., Mark, D. B., Rozema, T., Nahin, R. L. et al. 2012 Jan Design of the Trial to Assess Chelation Therapy (TACT). *Am. Heart J.* 163(1), 7–12. Available from: http://www.pubmedcentral.nih.gov/articlerender.fcgi?artid=3243954&tool=pmcentrez&rendertype=abstract
41. Lamas, G. A., Boineau, R., Goertz, C., Mark, D. B., Rosenberg, Y., Stylianou, M. et al. 2014 Apr EDTA chelation therapy alone and in combination with oral high-dose multivitamins and minerals for coronary disease: The factorial group results of the Trial to Assess Chelation Therapy. *Am. Heart J.* (Iv). Available from: http://linkinghub.elsevier.com/retrieve/pii/S0002870314001501
42. Kaufmann, P. A., Gnecchi-Ruscone, T., di Terlizzi, M., Schafers, K. P., Luscher, T. F. and Camici, P. G. 2000 Sep 12 Coronary heart disease in smokers: Vitamin C restores coronary microcirculatory function. *Circulation.* 102(11), 1233–8. Available from: http://circ.ahajournals.org/cgi/doi/10.1161/01.CIR.102.11.1233
43. Tveden-Nyborg, P. and Lykkesfeldt, J. 2013 Dec 10 Does vitamin C deficiency increase lifestyle-associated vascular disease progression? Evidence based on experimental and clinical studies. *Antioxid. Redox. Signal.* 19(17), 2084–104. Available from: http://www.ncbi.nlm.nih.gov/pubmed/23642093
44. Ali, A. and McCarthy, P. L. 2014 Dec 1 Complementary and integrative methods in fibromyalgia. *Pediatr. Rev.* 35(12), 510–8. Available from: http://pedsinreview.aappublications.org/cgi/doi/10.1542/pir.35-12-510
45. Suh, S.-Y., Bae, W. K., Ahn, H.-Y., Choi, S.-E., Jung, G.-C. and Yeom, C. H. 2012 Jan Intravenous vitamin C administration reduces fatigue in office workers: A double-blind randomized controlled trial. *Nutr. J.* 11(1), 7. Available from: http://www.pubmedcentral.nih.gov/articlerender.fcgi?artid=3273429&tool=pmcentrez&rendertype=abstract
46. Biesalski, H. K. and McGregor, G. P. Antioxidant therapy in critical care—Is the microcirculation the primary target? *Crit. Care. Med.* 2007. 35(9 Suppl), S577–83.
47. Marik, P. E., Khangoora, V., Rivera, R., Hooper, M. H. and Catravas, J. 2016 Hydrocortisone, Vitamin C and thiamine for the treatment of severe sepsis and septic shock: A retrospective before-after study. *Chest.* Available from: http://linkinghub.elsevier.com/retrieve/pii/S0012369216625643
48. Fisher, B. J., Kraskauskas, D., Martin, E. J., Farkas, D., Puri, P., Massey, H. D. et al. 2013 Aug 5 Attenuation of sepsis-induced organ injury in mice by vitamin C. *JPEN J. Parenter. Enteral. Nutr.* Available from: http://www.ncbi.nlm.nih.gov/pubmed/23917525
49. Fisher, B. J., Kraskauskas, D., Martin, E. J., Farkas, D., Wegelin, J. A., Brophy, D. et al. 2012 Jul 1 Mechanisms of attenuation of abdominal sepsis induced acute lung injury by ascorbic acid. *Am. J. Physiol. Lung Cell Mol. Physiol.* 303(1), 20–32. Available from: http://www.ncbi.nlm.nih.gov/pubmed/22523283
50. Hemilä, H. and Chalker, E. 2019 Vitamin C can shorten the length of stay in the ICU: A meta-analysis. *Nutrients* 11(4), 708. Available from: https://www.mdpi.com/2072-6643/11/4/708
51. Kositsawat, J. and Freeman, V. L. 2011 Dec Vitamin C and A1c relationship in the National Health and Nutrition Examination Survey (NHANES) 2003–2006. *J. Am. Coll. Nutr.* 30(6), 477–83. Available from: http://www.ncbi.nlm.nih.gov/pubmed/22331682
52. Casillas, S., Pomerantz, A., Surani, S. and Varon, J. 2018 Role of vitamin C in diabetic ketoacidosis: Is it ready for prime time? *World J. Diabetes.* 9(12), 206–8.
53. Traber, M. G., Buettner, G. R. and Bruno, R. S. 2019 The relationship between vitamin C status, the gut-liver axis, and metabolic syndrome. *Redox. Biol.* 21(October 2018), 101091. Available from: https://doi.org/10.1016/j.redox.2018.101091
54. Qutob, S., Dixon, S. J. and Wilson, J. X. 1998 Insulin stimulates vitamin C recycling and ascorbate accumulation in osteoblastic cells. *Endocrinology.* 139(1), 51–6.
55. Du, J., Cullen, J. J. and Buettner, G. R. 2012 Dec Ascorbic acid: Chemistry, biology and the treatment of cancer. *Biochim. Biophys. Acta.* 1826(12), 443–57. Available from: http://www.ncbi.nlm.nih.gov/pubmed/22728050

56. Riordan, H. D., Riordan, N. H., Jackson, J. A., Casciari, J. J., Hunninghake, R., González, M. J. et al. 2004 Jun Intravenous vitamin C as a chemotherapy agent: A report on clinical cases. *P. R. Health Sci. J.* 23(2), 115–8. Available from: http://prhsj.rcm.upr.edu/index.php/prhsj/article/view/469/332
57. Vollbracht, C., Schneider, B., Leendert, V., Weiss, G., Auerbach, L. and Beuth, J. 2011 Intravenous vitamin C administration improves quality of life in breast cancer patients during chemo-/radiotherapy and aftercare: Results of a retrospective, multicentre, epidemiological cohort study in Germany. *In Vivo.* 25(6), 983–90. Available from: http://www.ncbi.nlm.nih.gov/pubmed/22021693
58. Chen, P., Yu, J., Chalmers, B., Drisko, J., Yang, J., Li, B. et al. 2012 Apr Pharmacological ascorbate induces cytotoxicity in prostate cancer cells through ATP depletion and induction of autophagy. *Anticancer Drugs.* 23(4), 437–44. Available from: http://www.ncbi.nlm.nih.gov/pubmed/22205155
59. Drisko, J. A., Chapman, J. and Hunter, V. J. 2003 Apr The use of antioxidants with first-line chemotherapy in two cases of ovarian cancer. *J. Am. Coll. Nutr.* 22(2), 118–23. Available from: http://www.ncbi.nlm.nih.gov/pubmed/12672707
60. Carr, A. C. and McCall, C. The role of vitamin C in the treatment of pain: New insights. *J. Transl. Med.* 2017. 15(1), 77. Available from: http://translational-medicine.biomedcentral.com/articles/10.1186/s12967-017-1179-7
61. Carr, A. C., Vissers, M. C. M. and Cook, J. S. 2014 The effect of intravenous vitamin C on cancer- and chemotherapy-related fatigue and quality of life. *Front Oncol.* 4(October), 283. Available from: http://www.pubmedcentral.nih.gov/articlerender.fcgi?artid=4199254&tool=pmcentrez&rendertype=abstract
62. Riordan, N. H., Riordan, H. D. and Casciari, J. P., Ph D. 2000 Clinical and experimental experiences with intravenous vitamin C. *J. Orthomol. Med.* 15(4), 201–13.
63. Mikirova, N., Casciari, J., Rogers, A. and Taylor, P. 2012 Jan Effect of high-dose intravenous vitamin C on inflammation in cancer patients. *J. Transl. Med.* 10(1), 189. Available from: http://www.pubmedcentral.nih.gov/articlerender.fcgi?artid=3480897&tool=pmcentrez&rendertype=abstract
64. González, M. J., Miranda-Massari, J. R., Mora, E. M., Guzmán, A., Riordan, N. H., Riordan, H. D. et al. 2005 Mar Orthomolecular oncology review: Ascorbic acid and cancer 25 years later. *Integr. Cancer Ther.* 4(1), 32–44. Available from: http://ict.sagepub.com/content/4/1/32.short
65. Cameron, E. and Pauling, L. 1978 Sep Supplemental ascorbate in the supportive treatment of cancer: Reevaluation of prolongation of survival times in terminal human cancer. *Proc. Natl. Acad. Sci. USA* 75(9), 4538–42. Available from: http://www.pubmedcentral.nih.gov/articlerender.fcgi?artid=336151&tool=pmcentrez&rendertype=abstract
66. Creagan, E. T., Moertel, C. G., O'Fallon, J. R., Schutt, A. J., O'Connell, M. J., Rubin, J. et al. 1979 Sep 27 [cited 2013 Mar 22] Failure of high-dose vitamin C (ascorbic acid) therapy to benefit patients with advanced cancer. A controlled trial. *N. Engl. J. Med.* 301(13), 687–90. Available from: http://ukpmc.ac.uk/abstract/MED/384241
67. Drisko, J. A., Serrano, O. K., Spruce, L. R., Chen, Q. and Levine M. Treatment of pancreatic cancer with intravenous vitamin C. *Anticancer Drugs.* 2018;(March), 1. Available from: http://insights.ovid.com/crossref?an=00001813-900000000-98852
68. Dusing, R. W., Drisko, J. A., Grado, G. G., Levine, M., Holzbeierlein, J. M. and Van Veldhuizen, P. 2011 Aug [cited 2012 Oct 29] Prostate imaging modalities that can be used for complementary and alternative medicine clinical studies. *Urol. Clin. North. Am.* 38(3), 343–57. Available from: http://www.ncbi.nlm.nih.gov/pubmed/21798397
69. Doskey, C. M., Buranasudja, V., Wagner, B. A., Wilkes, J. G., Du, J., Cullen, J. J. et al. 2016 Tumor cells have decreased ability to metabolize H_2O_2: Implications for pharmacological ascorbate in cancer therapy. *Redox. Bio.* 10(October), 274–84. Available from: http://linkinghub.elsevier.com/retrieve/pii/S2213231716302634
70. Ngo, B., Van Riper, J. M., Cantley, L. C. and Yun, J. 2019 Targeting cancer vulnerabilities with high-dose vitamin C. *Nat. Rev. Cancer.* Available from: http://www.nature.com/articles/s41568-019-0135-7
71. Carr, A. 2019 Vitamin C Symposium 2019—"Vitamin C for cancer and infection: From bench to bedside." *Proceedings.* 5(1), 3. Available from: https://www.mdpi.com/2504-3900/5/1/3

72. van Gorkom, G., Klein Wolterink, R., Van Elssen, C., Wieten, L., Germeraad, W. and Bos, G. 2018 Influence of vitamin C on lymphocytes: An overview. *Antioxidants*. 7(3), 41.
73. Wilson, M. K., Baguley, B. C., Wall, C., Jameson, M. B. and Findlay, M. P. 2014 Review of high-dose intravenous vitamin C as an anticancer agent. *Asia Pac. J. Clin. Oncol.* 10, 22–37.
74. Ohno, S., Ohno, Y. and Suzuki, N. 2009 Mar High-dose vitamin C (ascorbic acid) therapy in the treatment of patients with advanced cancer. *Anticancer*. 29(3), 809–15. Available from: http://ar.iiarjournals.org/content/29/3/809.short
75. Gonzalez, M. J., Miranda-Massari, J. R., Duconge, J. and Berdiel, M. J. Increasing the effectiveness of intravenous vitamin C as an anticancer agent. *J. Orthomol. Med.* 2015. 30(1), 45–50.
76. Ong, C., Article, R. and Ong, C. 2019 High dose vitamin C and low dose chemo treatment. *J. Cancer Sci.* 5(1), 1–4.
77. Use, I., Padayatty, S. J., Sun, H., Wang, Y., Riordan, H. D., Hewitt, S. M. et al. 2004 Brief communication vitamin C pharmacokinetics : Implications for oral and. *Ann. Intern. Med.* 140(7), 533–8.
78. Wagner, B. A., Witmer, J. R., van't Erve, T. J., Buettner, G. R., Van't Erve, T. J., Buettner, G. R. et al. 2013 Jan [cited 2013 Jan 30] An assay for the rate of removal of extracellular hydrogen peroxide by cells. *Redox. Biol.* 1(1), 210–7. Available from: http://www.pubmedcentral.nih.gov/articlerender.fcgi?artid=3736862&tool=pmcentrez&rendertype=abstract
79. Oberley, L. W., Buettner, G. R. and Bueftner, G. R. 1979 Role of superoxide dismutase in cancer: A review. *Cancer Res.* 39(4), 1141–9.
80. Schoenfeld, J. D., Sibenaller, Z. A., Mapuskar, K. A., Wagner, B. A., Cramer-Morales, K. L., Furqan, M. et al. $O_2\cdot$- and H_2O_2-Mediated disruption of Fe metabolism causes the differential susceptibility of NSCLC and GBM cancer cells to pharmacological ascorbate. *Cancer Cell*. 2017. 31(4), 1–14. Available from: http://dx.doi.org/10.1016/j.ccell.2017.02.018
81. Alschuler, L. N. and Gazella, K. A. 2010 *The Definitive Guide to Cancer*, 3rd ed. InnoVision Health Media; Celestial Arts/Random House. 453 p.
82. Gyllenhaal, C., Alschuler, L., Rubin, D., Kranz, S., Roddy, G. D. and Block, K. I. 2008 Jun [cited 2013 Jan 24] Pancreatic cancer. *Integr. Cancer Ther.* 7(2), 103–13. Available from: http://www.ncbi.nlm.nih.gov/pubmed/18505898
83. Carr, A. C., Vissers, M. C., M. and Cook, J. 2014 Relief from cancer chemotherapy side effects with pharmacologic vitamin C. *J. New Zeal. Med. Assoc. NZMJ*. 24(127), 66–70. Available from: http://journal.nzma.org.nz/journal/127-1388/5964/
84. Riordan, N. H., Riordan, H. D., Meng, X., Li, Y. and Jackson, J. A. 1995 Mar [cited 2012 Nov 12] Intravenous ascorbate as a tumor cytotoxic chemotherapeutic agent. *Med. Hypotheses*. 44(3), 207–13. Available from: http://www.ncbi.nlm.nih.gov/pubmed/7609676
85. Riordan, H. D., Hunninghake, R. B., Riordan, N. H., Jackson, J. J., Meng, X., Taylor, P. et al. 2003 Sep Intravenous ascorbic acid: Protocol for its application and use. *P. R. Health Sci. J.* 22(3), 287–90. Available from: http://www.ncbi.nlm.nih.gov/pubmed/14619456
86. Ma, Y., Sullivan, G., Schrick, E., Choi, I. A convenient method for measuring blood ascorbate concentrations in patients receiving high-dose intravenous ascorbate. *J. Am. Coll Nutr.* 2013 Jan. 32(3), 187–93. Available from: http://www.pubmedcentral.nih.gov/articlerender.fcgi?artid=3725640&tool=pmcentrez&rendertype=abstract
87. Laurence, L. 2018 Brunton. In *Goodman & Gilman's: The Pharmacological Basis of Therapeutics* (Randa Hilal-Dandan, B. C. K., eds.), 13th ed, McGraw-Hill Education, New York, NY.
88. Sager, J. E., Yu, J., Ragueneau-Majlessi, I. and Isoherranen, N. 2015 Minireview Physiologically Based Pharmacokinetic (PBPK) modeling and simulation approaches: A systematic review of published models, applications, and model verifications. *Drug. Metab. Dispos.* 43(11), 1823–37.
89. Padayatty, S. J. and Levine, M. 2001 Feb 6 New insights into the physiology and pharmacology of vitamin C. *CMAJ*. 164(3), 353–5. Available from: http://www.pubmedcentral.nih.gov/articlerender.fcgi?artid=80729&tool=pmcentrez&rendertype=abstract

INDEX

C

D

I

J

K

L

M

R

S

T

U

V

W

X

Y

Z